LINES ACROSS EUROPE:
NATURE AND EXTENT OF COCAINE USE IN
BARCELONA, ROTTERDAM AND TURIN

LINES ACROSS EUROPE

NATURE AND EXTENT OF COCAINE USE IN BARCELONA, ROTTERDAM AND TURIN

B. Bieleman
A. Díaz
G. Merlo
Ch.D. Kaplan
(editors)

SWETS & ZEITLINGER

AMSTERDAM ∎ LISSE ∎ BERWYN, PA

∎ ACADEMIC PUBLISHING DIVISION ∎

Library of Congress Cataloging-in-Publication Data

(applied for)

Cip-gegevens Koninklijke Bibliotheek, Den Haag

Lines

Lines across Europe : nature and extent of cocaine use in Barcelona, Rotterdam and Turin /
B. Bieleman ... [et al.]. - Amsterdam [etc.] : Swets & Zeitlinger
Met lit. opg.
ISBN 90-265-1347-X
NUGI 735
Trefw.: druggebruik ; Barcelona / druggebruik ; Rotterdam / druggebruik ; Turijn.

Cover design: Rob Molthoff
Lay-out and graphics: E. de Bie
English editing: M. Marggraf-Lavery
Cover printed in the Netherlands by Casparie, IJsselstein
Printed in the Netherlands by Offsetdrukkerij Kanters B.V., Alblasserdam

ISBN 90 265 1347 X
NUGI 735

TABLE OF CONTENTS

Appendices

FOREWORD

The management of problems relating to the quality of contemporary life, in this case the question of cocaine use, requires a coordinated approach by politicians, researchers and social scientists. In the past, each group worked in isolation in their respective townhall or ivory tower. Research has now become so complex and politically sensitive that neither research scientists nor politicians have the ability to, on their own, produce results which provide meaningful answers to the questions posed by society. New research models involving the direct and daily input of both competencies need to be created. These models will breathe new life into the emergent field of policy science.

This book represents a new model of operating in the field of public mental health. It involved the input of scientists, policy planners and political decision makers. The study broke new ground in drug policy research. An entirely new synthesis of methods, results, conclusions and recommendations has emerged which will significantly influence the way people think about cocaine. The European Commission and the City Councils of Rotterdam, Barcelona and Turin showed great foresight in their willingness to support the study. Their work and the efforts of the researchers will be well rewarded if this book stimulates a debate throughout Europe on the nature and extent of cocaine and the formation of an appropriate policy.

This book is divided into three sections. Part I gives the background of the research project. This includes the Introduction, Legal background, State of the art, and Methodology. Part II outlines the results of the empirical research with separate chapters devoted to the General characteristics, Typologies of cocaine users and the Spread, dispersion and extent of cocaine use in each of the three cities. Part III describes the practical outcome of these results. The Summary and conclusions are followed by Recommendations for future work. After the Literature list and the Glossary come the Appendices describing the gestation of the project, acknowledgements, an explanation of star sampling, out-degree and multilevel analysis, the patterns of cocaine use, and the tables of characteristics of cocaine users.

A project which involves three different countries and three languages incurs a special difficulty of its own, the question of how to communicate. The Dutch, Italian and Spanish researchers, policy planners and political decision makers opted for the use of English. This meant that the onus of expressing complex ideas in a foreign language was shared by all. Great credit goes to all those involved, communication has been excellent throughout the project. For the names and functions of contributors see appendices A and B. Edgar de Bie and Cilia ten Den, two of the Rotterdam researchers, deserve special mention for their efforts in compiling and collating the huge amount of information coming in from the three cities into the working chapters. The next link in the chain was Mary Marggraf-Lavery who read each chapter and rendered the complex and technical ideas into an English that is intelligible to the general public. Last, but far from least, we wish to thank the cocaine users who discussed their experiences with us in lengthy interviews. Without their frank and enthusiastic cooperation, this book would not have been possible.

B. Bieleman
A. Díaz
G. Merlo
Ch. D. Kaplan

PART I

OVERVIEW

B. Bieleman, C. ten Den, E. de Bie, Ch. D. Kaplan

CHAPTER 1

INTRODUCTION

Cocaine is a topic receiving increasing attention in Europe. There is accumulating evidence that cocaine is currently available in all Member States of the European Community. The nature and extent of cocaine use is still largely unknown, however. The fact that more public attention is being paid to cocaine may be traced to a number of sources: the worsening problem of poly-drug use among opiate addicts; the reports of cocaine use in nightlife spots in European cities; the association of cocaine with glamour and successful careers; the sensational reporting of the international 'war on cocaine' being waged primarily by the United States against the organized criminals of South American cartels; and the general political climate which places drugs on the agendas of European politicians.

The increasing attention of the public to the cocaine issue has been paralleled by a scientific and clinical response to a perceived cocaine epidemic. Thus, Budney et al (1992) searched the Index Medicus from 1966 to 1990 to chart cocaine citations in the health and medical science literature. They found a dramatic increase in cocaine citations between 1983 and 1990. They compared this growth to the explosion of knowledge about cannabis and heroin a decade earlier. This increase in general scientific concern goes together with increased activities by political bodies responsible for policy-oriented research. For example, the Council of Europe Pompidou Group which is working in the field of drug epidemiology gives a high priority to the development of methods for studying cocaine use. (Pompidou Group 1992).

In the context of this growing public and scientific concern, this book represents a unique and coordinated effort of a multidisciplinary team of scientists, urban policy planners and political decision makers to increase public knowledge and awareness of the real extent and nature of cocaine use in Europe. It has often been said in terms of new drug epidemiological trends, Europe follows developments in the United States with a five to ten year lag. The question is whether this is a fact or a dangerous myth used to incite specific political interests and ambitions. This book presents the results of an intensive study of cocaine use in the cities of Rotterdam, Barcelona and Turin. The study was inspired by the desire to provide a rational and politically relevant contribution to the assessment of, and reduction in, the harm associated with the drug that has currently captured the imagination of Europe. Cocaine appears to be a drug with many facets rooted in all levels in society. The aim of the study was to identify the true nature of cocaine use in Europe. Once this is known, it can be properly evaluated. The method used was to establish contact with a broad based sample of cocaine users in three culturally diverse European communities.

1.1 Local situation

The three cities conducted intensive local studies on the nature and extent of cocaine in 1992 (Diaz et al 1992, Intraval 1992, Merlo et al 1992). This book is based on the results of these three studies. Each city has its own special character within the framework of Europe.

Rotterdam is the second largest city in the Netherlands. With its population of 575,000 in 1991, it is considerably smaller than Barcelona and Turin. However, as Rotterdam is also responsible for drug addiction help in the surrounding Rijnmond region, the city has a service population of more than a million people. Rotterdam has a rich maritime history and is the largest port in the world. More than 20,000 containers pass through the port each day. It is highly likely that some of these containers have cocaine hidden in their legitimate cargo. Rotterdam is a North Sea port at the mouth of the River Rhine, a major European waterway. To the south are the massive flood defences known as the Delta Works, while to the north are the cities of The Hague and Amsterdam. The river divides Rotterdam into two distinct sections. The centre of the city, which was rebuilt after the Second World War, is on the north bank. It is characterized by the integration of modern and traditional architecture. Whereas Amsterdam is seen as the cultural centre of the Netherlands, Rotterdam is thought of as a city of blue-collar workers. The industrial sector, particularly the port and oil refinery facilities, provides jobs for many Rotterdamers. The motto of the city reflects its hard-working, no-nonsense character: 'strength through struggle'.

Until the mid-1970s, problems with drug taking were virtually unknown in Rotterdam. From 1976 onwards, the nuisance created by drug users and drug dealers increased sharply. At the end of the decade, the Rotterdam City Council published its first reports on the drug problem. These reports clearly indicate that the city did not view addiction as a disease but as a social problem. The problems experienced by the addicts were to be dealt with primarily by the social services and treatment agencies, while the problems caused by addicts were considered to be primarily the responsibility of the police and the criminal justice system. The City Council has never had illusions about finding a quick solution to the drug problem. Efforts have always been focused on containing addict behaviour within the social dimension and, where possible, eliminating it. The objectives of the Rotterdam drug policy are: to prevent the onset of drug use by non-users; to provide assistance to those who want to kick the drug habit; to provide (as far as those who want to continue taking drugs are concerned) low threshold assistance programmes which enable addicts to live a socially acceptable life; and to combat drug dealing and to decrease drug related nuisance problems and drug-related crime.

Barcelona, the capital of Catalonia, is Spain's second largest city. It is a port, in the north-west of the country, 150 kilometres from the Pyrenees and the French border. In 1991, the population of Barcelona was 1,643,542. Including the surrounding suburbs, this figure rises to 3,700,000 inhabitants. Over the last century three international events featuring the city have entailed important goals in its economic and urban development: the 1888 Universal Exhibition; the 1929 International Exhibition; and the 1992 Olympic Games. In the course of this century, Barcelona has become one of the most powerful industrial centres of Spain. Industry has progressively moved outside the city. Currently, it is the development of the service sector which is Barcelona's main economic feature. A remarkable process of attractive urban planning has accompanied the expansion of the city. Over the last ten years, Barcelona has developed into an international centre of the arts.

In Spain, drugs first appeared in the period 1973-1978. Cannabis and LSD was used in cross-cultural groups in the youth population. Hedonistic and poly-drug habits, including the use of heroin, were increasingly evident. The social and health care services were taken by surprise when, in the mid-1970s, the first request came for treatment from heroin addicts. The existing services were geared only to deal with alcohol addiction. To cope with the new problems, a number of centres for drug addiction treatment were opened. In 1980, the city of Barcelona set up the Integral Care Programme for Drug Addiction. In 1985, in line with the National Plan for Drugs, the existing services and resources were integrated in the Network of Drug Addiction Care Centres of the Catalonian Regional Government. In 1988, a Municipal Plan of Action against Drugs was agreed on. This aimed to: reduce the risk of an individual becoming addicted; reduce the supply and promotion of legal drugs; improve health and social care facilities for drug addicts and make them more accessible; establish a territorial setting for the treatment and social rehabilitation of addicts and their families; coordinate all network resources; and obtain information about drug problems in the city and the care of users.

Turin, an inland city in the north-west of Italy, had 979,839 inhabitants in 1991. Including the suburbs, the population amounted to around 1,800,000. Over the last thirty years the population in the suburbs has been steadily increasing and has become relatively larger than the city population. Turin, like Rotterdam, is a blue-collar industrial city. In spite of the recent expansion of the service sector, the city continues to be a centre of industry, or more precisely, industrial production. FIAT, the Italian motor industry giant, has been the most influential factor on the development of the Turin economy. Turin is characterized by a highly organized system of business contacts. These relationships influence the structure of the whole of society, including the cultural sector (Bagnasco 1986).

In Turin, as in Rotterdam and Barcelona, it was in the early 1970s that the problem of hard drugs spread beyond narrow and exclusive circles. The social use of heroin became highly visible in certain small groups. After 1975, there was a rapid increase of this phenomenon among groups of young people who appeared to be increasingly alienated from the rest of society. In 1980/81 a massive spread of the use of cocaine combined with heroin was observed, but, in no time, the phenomenon disappeared. This sudden cessation was triggered by the fear of drug addicts that the mixing of cocaine with heroin was responsible for the serious symptoms of mental illness in many people. In the second half of the Eighties the spread of heroin use was reduced. In 1991, the number of heroin addicts in Turin was estimated at 7,500. Heroin use seemed to have become socially integrated with less visibility of addictive behaviour and a degree of adaptation to the problem by the general population. As heroin became less visible, there were the first signs of the spread of cocaine in different sectors of society.

1.2　Problem formulation

Compared to heroin, cocaine is a drug with many facets. Cocaine is used not only by so-called captive groups such as opiate addicts, but also by other categories which are harder to identify. Furthermore, the true addiction risks of cocaine are unclear and little is known about its connections with deviant or criminal behaviour. Specific cocaine studies are relatively recent. In contrast to research among opiate addicts, there is no tried and proven method of study, nor is there a broad basis of knowledge for comparing findings. In order to gain more insight into the phenomenon, the cities of Rotterdam, Barcelona and Turin decided to carry out extensive study into cocaine use among their populations. The studies have been conducted more or less simultaneously with the researchers and supporting officials working in alignment. The same approach, based on the Intraval research design, has been used in each city (Intraval 1990). The key question was:

What is the nature and extent of cocaine use in Rotterdam, Barcelona and Turin?

This can be divided into four subject areas:
1. The nature of cocaine use.
2. The extent and distribution of cocaine use.
3. Prevention and intervention possibilities.
4. The differences and similarities (i.e. the comparative analyses) between Rotterdam, Barcelona and Turin.

The nature of cocaine use

In studying the nature of cocaine use, attention will be paid to the following aspects.

a. User/drug relationship

Cocaine is attractive due to its status and (alleged) qualities as a harmless but stimulating glamour drug. This fits the achievement mentality of present day society. Other important aspects are the (social) circumstances of the individual user when he takes cocaine for the first time and in subsequent periods. Furthermore, the situations in which people use cocaine need to be studied. It is possible that this factor will, to some extent, determine the substances with which cocaine is combined, as well as the method and frequency of use. These aspects may influence the user-drug relationship and the level of addiction. An important question is the relationship between problematic and non-problematic use.

b. Social categories of users

Cocaine use is frequently observed in circles far removed from the familiar heroin addict groups. This means, in practice, that use is often related to less tangible groups (non-captive groups), the so-called hidden populations. Cocaine is often associated with fast and big spending and with rapid and copious consumption, and with social night life such as discotheques, trendy bars, and house parties. The real difference with heroin relates to the identity of the user. The use of heroin implies a certain lifestyle. Cocaine, on the other hand, is used as a sideline. People take cocaine but their life is centred on other things.

c. Connections with deviant and criminal behaviour

The connections with deviant and criminal behaviour may be listed as: drug trafficking, acquisition crimes, violent offenses; combined effects of poly-drug use (including alcohol), group vandalism and visual nuisance. The type of social nuisance involved is a matter of discussion. At the level of the user a distinction can be made between criminal activity surrounding small-scale dealing, income-generating crime, and other crime-reinforcing effects of cocaine.

The extent and distribution of cocaine use

Cocaine use often occurs in hidden populations. Knowledge is lacking, most users cannot be identified, and they rarely come into contact with police or care centres. It is therefore very difficult to estimate the extent of cocaine use. In this research project, we have tried to develop new methods to estimate the extent of cocaine use, e.g. a combination of snowball sampling and network analysis. In addition, the network analysis can also provide insight into the distribution and dispersion of cocaine use in the three cities.

Prevention and intervention possibilities

Analysis of the research data may reveal that cocaine use results in problems for (groups of) users and/or society. If so, a policy of prevention and intervention will need to be developed. The research material may provide various leads for this. One could think of such areas as: the availability of cocaine; information on the negative aspects; shattering the myths surrounding the drug; and intervention in individual cocaine use, careers and lifestyles (particularly at strategic moments).

CHAPTER 2

LEGAL BACKGROUND

Cocaine is an illegal drug. As a controlled substance, it is generally considered in the class of heroin as a so-called hard drug. Therefore, in order to understand the social context of cocaine use, a general overview of the origins and structure of the contemporary legal framework controlling cocaine is necessary. The origin of this framework can be found in the development of the modern technocratic welfare state and social movements advocating crusades to save modern mankind from degradation through the creation of new forms of moral conscience (Escohotado 1989). Bernard Leroy (1992) in a comprehensive comparative study of the legal structure and judicial practice relating to drugs in the European Community, elaborates this historical process in the Twentieth Century. He argues that since the beginning of the century, drug use has, at successive times, been perceived as a sin, a disease, and a lifestyle. Each perceptual orientation carried its own legal consequences. Perceiving drug and alcohol use as a sin, in the first two decades of this century, reinforced the Nineteenth Century moralistic campaigns aimed at changing attitudes and the drug use patterns of the population.

The failure of these voluntary moral appeals to change attitudes to drugs and alcohol stimulated a search for more effective measures. The law was seen as a potentially effective instrument in curbing drug and alcohol demand. International diplomats and state administrators joined with moral crusaders in integrating the demands of prohibitionist social movements into the legal apparatus of the state (Stein 1985). The International Treaty of The Hague (1912) restricted opium, morphine, cocaine and their salts (and any new derivative from them) to medical and legitimate uses and limited their export. The tenets of The Hague Treaty (later incorporated in the Treaty of Versailles of 1919) were subscribed to by the majority of the nations of the world including Italy, the Netherlands and Spain (see Escohotado 1989). These international developments created a legal framework for a new world order relating to drug control which can be characterized by an uneasy consensus between prohibitionist and experimentalist expectancies (Kaplan 1984). The United States was the first country in which opiate and cocaine users were taken to court in large numbers according to the mandate provided by The Hague Treaty. National and local laws were enacted and constitutionally tested to support prohibition of drugs and alcohol. Later this wave of prohibition spread to Europe for drugs such as opium and cocaine. European prohibitionist laws appeared between the First and Second World Wars.

As with earlier moral appeals, the simple punitive legal measures which continued after the Second World War failed to effectively curb drug and alcohol demand. A new shift in emphasis was seen, this time towards an orientation that defined the social and relational problems tied to drug and alcohol use as medical (psychiatric) diseases. With the further increase of drug addiction that developed in the late Sixties, a change occurred in the scale of the problem. Approaches based on narrow legalistic and medical thinking were seen to be inadequate. New professions were drawn into the problem; most notably the social sciences. With the ascent of the social sciences, the approach to drug demand also changed to one of deviant behaviour and lifestyle adaptations. The result of this redefinition has been to reconsider drug and alcohol use as a part of a collective group and cultural life and to approach the problem from the point of view of public policy management rather than based on a sick or criminal image of the drug user.

The legal approaches adopted by Italy, the Netherlands and Spain have their roots in these new social scientific definitions of the drug problem. In dealing with demand and supply, legislation in all three countries concentrates on the supply side. In terms of demand, these countries' legislation is among the most tolerant in Europe but there are significant indications of a new shift to increasing legal intolerance. A systematic comparison of the legislation reveals differences in the local situations of each country and in the perception of the role of criminal law. Whereas Italian legislation, especially since the reforms of 1990, is comparatively less permissive, Dutch and Spanish legislation has depenalized personal drug use with the aim of harm reduction, an approach to drug control that places emphasis on the real harm that drugs effect on the individual user and society (Taradash 1991). Harm reduction is committed to developing drug policies that do not cause even more harmful consequences than are caused by the use of drugs themselves.

2.1 Legal history

Italy

The history of Italian legislation on drugs is characterized by swings back and forth from sympathetic understanding to repression. This corresponds to two distinct ideas: the drug addict as a sick person and the drug addict as a criminal. The first Italian laws were made in line with the The Hague Treaty. The 1923 Law referred to 'poisonous substances with narcotic effects' and emphasised the 'psychic changes' due to the 'abuse' of these substances. The 1934 law (No. 1145) placed the drug addict in the same category as the insane and included a procedure for admission to a mental hospital. The 1954 Law (No. 1204) prescribed heavy penalties (three to eight years imprisonment) not only for dealing but also for the possession of any listed substance. It was referred to as the 'two years for one gram' law because the possession of small quantities of drugs for personal use incurred a mandatory sentence of two years. This meant that the same sentence could be given to the small-scale user and the large-scale trafficker. Public opinion subsequently demanded the modification of this unfair legislation, so in the early Seventies the debate was reopened. In the meantime the opinion on how to deal with drug users had shifted from repression to medical treatment. The 1975 Law (No. 685) embodied a distinction between light and hard drug penalties, and between users and pushers. It established social-health services for the prevention, cure and rehabilitation of drug addicts. Punishment for the possession of moderate quantities was abolished. The question of a drug addict's liability was now entrusted to the decision of a judge. The 1980s, however, was a period of renewed prohibitionist and repressive attitudes. Discussion was characterized mainly by ideological disputes with little account being taken of technical scientific knowledge.

The most recent law (No. 662 and Consolidation Act 309/90) dates from 1990. Article 13 states that 'the personal use of drugs and psychotropic substances is forbidden'. Some legal opinions have defined this as a strict rule. The significance of this must be seen in the context of the fight against permissiveness, rather than in a technical-legal context. Earlier legislation already contained strictures against the illegality of drug-related behaviour but recognized that the drug user was not liable to punishment for possession of a moderate amount. It was up to the judge to decide what moderate meant. In the new law, a 'moderate amount' has been replaced by 'a fixed average dose' laid down by a Ministry of Health decree. Anyone caught in possession of an amount exceeding the limits of the decree may be charged with 'possession for dealing'. Charges that new legal regulations were unconstitutional resulted in the Constitutional Court insisting on a more cautious enforcement of the 1990 law. A wave of protest followed after many young people, imprisoned for possession of hashish or marijuana which exceeded the average daily dose, committed suicide. They had been automatically charged with drug dealing, though it was clear that they belonged to the category of consumers. Subsequently, the 1990 law was modified.

The measures of the new law do not deal with the ambiguity in determining the categories use, dealing, and trafficking. In practice, this ambiguity leaves much to the discretionary power of the judge. The amount of drugs one is allowed to possess in order not to be regarded as a criminal dealer is 0.15 grams for cocaine and 0.02 grams for crack cocaine. In the past, based on sentences of the Supreme Court of Appeal, the limit for cocaine was seven grams. Unless a user can prove (inversion of the burden of proof) that the dose in his possession is exclusively for personal use, he runs the risk of very heavy sanctions. The new law, furthermore, punishes certain drug-related offenses, such as taking drugs in a public place, with three to ten years' imprisonment. A single offence is sufficient to incur imprisonment. The owner of premises where drugs are taken also risks a prison sentence, or the closing of his premises for two to five years, even when he is not personally involved in using or supplying drugs. The penalties are even more severe in the case of hard drugs, particularly if an underage person is present. The consequences of such strict law enforcement at the drug user level has seriously hindered the drug assistance services which have invested years of work in building up an image of trust in the drug user population. Recent critics of the new laws have argued strongly that these new legal conditions make drug counselling and treatment impossible.

The Netherlands

The first Dutch drug legislation was enacted in 1919 after The Hague Convention of 1912. The first law, known as the Opium Act, has been modified and revised a number of times. The Netherlands is a signatory to the 1961 Single Convention on Narcotic Drugs and therefore has legal provisions for the punishment of activities involved in the possession and trafficking of listed drugs except if they are being used for scientific and medical purposes. The first legal actions against listed drugs in the current era was in the mid-Fifties when the Public Prosecutor requested (and obtained) stiff prison sentences (more than one year) for cannabis offenses. Convictions and sentences of more than one year continued through the mid-Sixties (Van de Wijngaart 1991, Van der Meulen 1969). This strategy of vigorous prosecution of cannabis offenses was seen as an effective deterrent to the perceived threat of an increase of cannabis use. By the late Sixties an increasing number of experts and officials criticized the use of criminal law as ineffective in preventing the spread of cannabis use.

A lively debate ensued through the mid-Seventies where critics and defenders of the use of criminal law battled for their interests. The net result of this public debate was the 1976 Amendment to the Opium Act. This revised Act is a 'compromise between outright prohibition and social integration of illegal drugs' (Leuw 1991). The amendment was based on a legal distinction between 'drugs presenting unacceptable risks' such as heroin, cocaine, amphetamines and LSD (referred to as Schedule I) from 'traditional hemp products' such as hashish and marijuana (Schedule II). Further prosecution guidelines were laid down subjecting the possession of illegal drugs with less severe penalties than international trafficking in these drugs. The decriminalization of cannabis has aimed at separating the markets and social worlds of soft and hard drugs. The spirit of the these amendments has been expressed by Van der Wijngaart (1991): 'The Act reflects the view of the Dutch government that criminal law plays only a minor part in preventing individual drug use. Although the risks to society must be taken into account, every possible effort must be made to ensure that drug users are not caused more harm by the criminal proceedings than by the use of the drug itself.'

In the Dutch government view the penal law is 'part of the Dutch drug policy framework that includes tolerance for nonconforming lifestyles, risk reduction in regard to the harmful health and social consequences of drug taking, and penal measures directed against illicit trafficking in hard drugs' (Leuw 1991). Therefore, the responsibility for implementing the Opium Act rests jointly with the Minister for Welfare, Health and Cultural Affairs and the Minister of Justice. The general trend of the revised act includes the following: a reduction of all penalties for Schedule II drugs; a reduction of penalties for possession of both Schedule I and Schedule II drugs (in the 1928 revision possession of Schedule I carried a maximum four year sentence, lowered to one year in the 1976 Act); a differentiation in maximum penalties for different aspects of drug trafficking; and an increase in the maximum penalties for trafficking of Schedule I drugs from four to 12 years.

A unique feature of Dutch law enforcement has been the exercise of the expediency principle. As far as trafficking in heroin and cocaine is concerned, the Netherlands resembles most other Western countries and operates in accordance with the international legal drug control order. However, in other areas of drug law enforcement, activities take place within a wider framework of local city policy relating to public health, public order and welfare. The expediency principle authorizes the public prosecutor to decide whether to initiate criminal proceedings in specific cases. These decisions are ultimately based on the political responsibility of the Minister of Justice. The guidelines issued by the Minister of Justice in 1976 comprise the basis of the most recent Opium Law reform. They empower the public prosecutor to set the priorities of police enforcement strategies. Derived from these guidelines have been some specific police norms. In a study of the prosecution policies of 1,042 drug offenses involving the public prosecutor in 1982 it was found that there was more leniency than the guidelines intended; e.g. 64% of the sentences requested in possession-for-trade cases were below the one year term set by the guidelines and 56% of the sentences requested for international trafficking were below the two year minimum (Rook and Essers 1987).

In The Netherlands there are no legal provisions for compulsory treatment of drug addicts under the drug law as exist in other European countries. Formally, compulsory treatment is only possible within the framework of forensic psychiatry in general. However, now that the number of drug addicts in some prisons is above 50% of the prison population, the Ministers of Justice and Health sent a joint memorandum to Parliament to urge for new creative measures to be taken to pressure addicts to undergo treatment as an alternative to prison (see Engelsman 1989, Van de Wijngaart 1991). To date, however, there is still no legal mechanism for compulsory treatment.

Spain

In Spain, drug use is neither prohibited nor the acquisition or possession of drugs with a view to use penalized. In coping with the supply and demand of drugs, Spanish legislation has always focused on the supply of drugs. Spain adheres (sometimes with relative delay) to the majority of international treaties concerning drug control (Del Castillo 1981). The first reference to drugs in the Spanish legislation appeared in the 1928 Penal Code (López-Muñiz 1980). In the articles of this code, drugs or narcotic traffic are considered a serious crime against public health. This specific legal criterion lasted for many years (Aranzadi 1945). The 1963 Penal Code, modified by the 1967 April law, can be considered as the actual legal precedent on which the rules concerning drugs are based (Del Castillo 1981).

When Spain ratified in 1966 the 1961 Single Convention Treaty on narcotics, it endorsed the policies drawn up by the United Nations (González et al 1989). In line with the aim of this treaty to restrict the illicit traffic of toxic drugs and narcotics, the Spanish legislature enacted the 1971 Law, later modified as Article 344 of the Penal Code. The law states: 'it is the responsibility of the Penal Code to prevent and punish any behaviour concerning the production, possession and trafficking of drugs as well punishing those who propagate its use ....' The new precept of the Code is intended to sanction two offenses: production of and trafficking in toxic substances and narcotics and their non-medical prescription or sale. Two important aspects are that the law does not define legally what is understood by 'toxic drugs and narcotics'. Jurisprudence has understood that those substances were the ones mentioned in the lists attached to the 1961 Single Treaty (González et al 1989). Secondly, the law does not punish the possession of drugs or narcotics in small amounts for a personal use. Nevertheless, the incrimination of possession for use remains unspecified. Furthermore, it needs to be recognized that in the application of the 1970 'Ley de Peligrosidad y Rehabilitación Social' (Law of Risk and Social Rehabilitation), 'addicts' have been declared as people in a 'dangerous state' and therefore subject to security measures. In practice this has meant that drug users were subject to a double, stricter, penal sanction.

The lack of specification of toxic substances, the severity of punishment and the wide range of sentencing, among other reasons, constituted the main criticisms of the Penal Code regulations. According to some authors, this criticism is well justified because the regulations did not respect the principle of legality (see González et al 1989). The negative consequences resulting from the application of the regulations of the Penal Code promoted the Urgent and Partial Reform of the Penal Code of 1983. This Reform modified Article 344 of the Penal Code. In the new draft, psychotropic substances were put on the same level as toxic drugs and narcotics. With this levelling, the agreements adopted in the Treaty of Vienna of 1971, which were ratified by Spain in 1973, were finally added. In the precept the legislation's will to differentiate hard drugs from soft drugs in the criminal prosecution of drug trafficking is expressed. As far as punishment is concerned, a distinction is made between 'a substance that can cause serious health damage' and cannabis and its derivatives. Another important aspect of the 1983 reform is the clear exclusion, in Article 344, of the possession for use as well as the reduction of prohibited acts of cultivating, producing, dealing and possessing drugs for the purpose of dealing. Consequently, the fact of giving cocaine or inciting its use cannot be punished unless it is also considered to be linked to specific acts of dealing and trafficking. Nevertheless, it must be mentioned that the legal criteria relative to the possession for private use alone are not always uniform (e.g. the legal difficulty of establishing the line between possession for one's own use or for dealing).

The Reform of 1983 was portrayed by some political sectors and part of the mass media as a decriminalization of drugs. There was a strong opposition to it from a campaign against public insecurity. The Reform was considered to have increased public insecurity and was blamed for the significant rise in crime (especially robbery) in 1983 and 1984. Likewise, in order to explain the subsequent hardening of the Penal Code (the so-called counter-reform of 1988), we have to look the way in which international pressures have affected Spain. In 1985, the United States declared a war on cocaine and agitated for a new international convention relating to drug trafficking. The effect of this was felt in Spain. It coincided with a renewed public demand, widely publicised in the mass media, demanding political intervention to counter dealing by relatives of drug addicts. Consequently, the Penal Code relating to illegal drug dealing was revised in 1988. This revision is still in force in 1992. The intention of the Law is to pursue drug dealing with increased efficiency by participating in the coordinated action of the various social institutions engaged in the 'Plan Nacional Sobre Drogas' (National Drugs Plan) established in 1985.

An important item of these most recent legal changes is the increase in the restriction of freedom in certain cases. Initially intended to punish drug dealing, in practice, it has meant harsher sentences for any persons convicted for any type of behaviour considered to be dealing. The law includes the possibility for a judicial authority to conditionally suspend the sentence of a drug addict arrested for crimes committed in order to support this dependency. To obtain this concession and avoid a two year sentence, the person concerned must prove that he has given up narcotics or that he has enroled in treatment

---

designed to achieve this. At the beginning of 1991 the mass media began discussing the administrative punishment of public drug use. A number of local authorities published advance 'Bandos Antidroga' (Edicts Against Drugs) according to which fines were levied on public drug use and the use of hypodermic syringes in public places. Finally, in February 1992, in line with a general hardening of the attitude towards drug dealing and consumption, the highly controversial Organic Law on the Protection of Public Safety ('Corcuera's Law') was promulgated.

2.2   Comparison of legislation

In systematically compared legislation relating to drugs in the European Community, Leroy (1992) has constructed a table cross-classifying each Member State according to several dimensions.

Table 2.1.   Synopsis of National Legislation for Italy, Netherlands and Spain (source: Leroy 1992)

	Italy (1990)	Netherlands (1928 and 1976)	Spain (1983 and 1988)
Product-classification	Distinction between cannabis and other drugs	Distinction between cannabis and other drugs	Distinction between cannabis and other drugs
Drug use	Prohibited and punished	Not an offence	Not an offence
Notification	Notification to health authority	None	Law about social dangerousness of drug addicts
Combined treatment-penalty procedure	Voluntary incentives compulsory	Voluntary incentives compulsory	Voluntary incentives compulsory
Possession with a view to use	1st, 2nd: suspension of driving licence; 3rd: prison and fine	SD: to 3 months HD: to 1 year	Not an offence
Drug possession	SD: 2 - 6 years HD: 4 - 15 years	To 2 years imprisonment	SD: 4 months to 4 years HD: 8 months to 8 years
Supplying drugs to users	4 - 15 years aggravat. 1/3* imprisonment	SD: to 2 years HD: to 8 years (national)	SD: 6 months to 6 years; HD: 6 - 14 years
Drug trafficking	4 - 15 years HD: 20 years minim. imprisonment	SD: to 4 years HD: to 12 years (international)	SD: 10 - 17 years* HD: 14 - 23 years*

HD: hard drugs;   SD: soft drugs;   *: aggravated penalties.

Table 2.1 extracts, up-dates and modifies the data for Italy, Netherlands and Spain for the purpose of comparative analysis.

Product classification

Italy, the Netherlands and Spain are among the countries which make a distinction between hard and soft drugs. Spain draws a distinction between substances 'which can cause serious damage to health' and cannabis and its derivatives. The Netherlands adopts a similar approach, identifying substances 'presenting an unacceptable risk' (such as heroin, cocaine, amphetamines and LSD) and cannabis derivatives. Italy divides the categories into highly addictive drugs (Schedule I and III of the Single Convention, 1961) and drugs with less addictive properties and cannabis (Schedules II and IV). In all three countries the penalties relating to hard drugs are more severe than for soft drugs. Cocaine is uniformly classified as a hard drug in all three countries.

Drug use

In Italy, the use of drugs in any form is prohibited and, above a prescribed amount, penalized as a criminal offence. Italian law allows for therapy orders, which are progressively more binding, on persons found in possession of drugs. Spain and the Netherlands are the only European countries in which drug use is not an offence. In Spain, legal repression of drug use is formally not possible. In the Netherlands, provision is made for prosecution for possession with a view to use. In practice, however, legal action is rarely taken. In Italy, the law prohibits cocaine use. In Spain and the Netherlands, cocaine use is regulated by informal social control.

Notification

In Italy, the administrative authorities must notify the health services when a drug user is detained by the police. The Drug Dependence Service is required to contact the addict and arrange a therapy programme with guaranteed anonymity. If the addict refuses treatment, the judge must be notified. In the Netherlands there are no legal provisions for notifying the health services of the arrest of an addict. In Spain, notification is possible under the provisions of the law relating to dangerous behaviour of drug addicts, but it is not uniformly applied.

Combined treatment-penalty procedure

Drug users facing criminal charges in Italy can suspend criminal procedures by requesting treatment. In the case of a first offence, if the treatment is successfully completed, charges may be dropped. Spanish law provides an incentive for addicts to undergo treatment in the form of a conditional suspension of prison sentence. In the Netherlands, treatment does not affect the legal situation of addicts though it may be included in the probationary measures applied at the time of sentencing. Detoxification treatment may be ordered instead of prison for users of hard drugs. In Italy and Spain provisions exist which allow legal authorities to take action on an individual's drug dependence. In both countries these provisions exist in addition to specific repressive measures. In Italy an emergency order is possible to force an addict to be treated before an actual offence has been committed. In addition, addicts on the third or subsequent possession charge may be forced into treatment. Noncompliance can result in a prison sentence or a fine. Spanish law contains a public protection provision for socially dangerous addicts allowing for their mandatory treatment and other restrictions. These measures may be taken before any offense has been committed. In the Netherlands, there are no provisions for therapy under the law. Compulsory detoxification orders are possible for users of hard drugs, but only under the terms of the general legal codes. There are no specific legal provisions for drug addicts. However, compulsory treatment is possible as an alternative to imprisonment. Thus, in all three countries, there are provisions for a combination of voluntary incentives and compulsory measures.

Possession with a view to use

In Spain, possession with a view to use is not an offence. In Italy, it can result in a prison sentence and a fine for a third offence. In the Netherlands, legal provisions allow for imprisonment for soft drugs for up to three months and for hard drugs up to a year. These provisions are rarely applied in practice, however.

Drug possession

In all three countries a distinction is made between possession of drugs for personal use and possession with a view supplying other users. In Italy the possession of drugs clearly for personal use earn sentences of two to six years for soft drugs and four to 15 years for hard drugs. In Spain the penalties are more lenient, with four months to four years for possession of soft drugs and eight months to eight years for hard drugs. In the Netherlands the maximum sentence for drug possession is two years imprisonment.

Drug supply

In all three countries supplying drugs to users is considered a serious offence. In Italy, punishment ranges from four to 15 years with graded sentencing. In Spain the sentences are lower: from six months to six years. For hard drugs the maximum can be extended to 14 years. In the Netherlands the maximum sentence is two years for soft drugs and eight years for hard drugs. In the Netherlands a distinction is made between national and international drugs trafficking and between soft and hard drugs. The penalties are more severe for international trafficking and for hard drugs trafficking. Spain adopts a similar sentencing pattern. The most severe penalties of the three countries are found in Italy: four to 15 years for soft drugs and 20 years minimum for trafficking in hard drugs.

CHAPTER 3

STATE OF THE ART

Comparatively little research has been done on cocaine and knowledge of the drug, particularly in Europe, is scarce. Now that specific studies are being undertaken, the body of knowledge is growing. In this chapter attention will be paid to the state of the art relating to cocaine in the areas of: supply and demand; prevalence; pharmacology and addiction; morbidity and mortality; and social risks and criminality. In the closing sections of the chapter several socio-cultural/historical studies will be discussed.

3.1 Supply

According to Interpol data, the total quantity of cocaine confiscated in Europe in 1978 (155 kg) was just enough to fill one suitcase. By 1988 the total quantity would have filled a standard ship container. Furthermore, 1988 was the first year in which more cocaine was seized in transit than heroin (5.3 tonnes cocaine compared to 2.2 tonnes heroin). The 1989 figure was even higher, 6.1 tonnes (Interpol 1990). The country in which the greatest quantities of cocaine are intercepted is Spain. Almost half of the cocaine seized in 1991 in 15 European countries, had been intercepted in Spain (7.5 tonnes in 1991). Furthermore, since 1988, small quantities of crack cocaine have been seized, 430 grams in 1991 (PNSD 1992). Next is the Netherlands, but the gap is shrinking. In 1991, 2.5 tonnes of cocaine was intercepted in the Netherlands (WVC 1992). Both the Netherlands and Spain appear to have an important transit function for Europe. Germany is placed third on the Interpol list, followed by France, Portugal and Italy (Interpol 1990). While these data have a certain value as an indirect indicator of cocaine use, they are not useful in estimating supply as they are often confounded by policy decisions which define the allocation of scarce police resources intended for supply interception.

At present, most cocaine seizures in Europe are made at the border cross-ings. There is much smuggling of large quantities of drugs in airplane cargoes or on board ships. Aircraft and ships provide an abundance of ideal conceal-ment locations which virtually defy detection. According to Jeschke (1987), the reason for the growth in cocaine supplies to Europe is the near saturation of the North American market. The resulting fall in prices, reinforced by the downward trend in the exchange rate of the dollar, seems to be an equally important factor. Jeschke (1987) concludes that it is unclear whether the buoyant supply in Western Europe is a response to consumer demand or a strategy to open a significant European market: 'Many indications, mainly the figures on new consumers known to the police, suggest that far more cocaine is being imported into Europe than is at present being consumed. On the other hand, the growing number of consumers would suggest an expanding market'. These trends are very difficult to monitor because the consumption of drugs, especially cocaine, usually takes place in private circuits and the social deviance of cocaine addicts, as distinct from heroin addicts, is slow to manifest itself.

Nonetheless, Western Europe has emerged as a homogenous, closed market for drugs in the last two decades especially for cocaine (Jeschke 1987, Lewis 1989). As Kaplan (1987) points out, in this period the European drugs market was restructured 'from basically an informal, almost barter system into a highly structured hierarchical system with elements of verticle and horizontal concentration'. He describes a dynamic from peer-oriented markets of dealer/ consumers to highly structured, organized criminal hierarchies where the drug user plays an entirely different role. The European market seems to increas-ingly resemble the American market, being equally attractive to international trafficking organizations. According to Jeschke (1987), the wholesale cocaine trade is in the hands of established groups. The organization in Latin America is paralleled by marketing organizations in Europe. Most of the profits are achieved on the basis of business calculations. The risk of losses, storage, etc., are factors included in the calculations. The couriers employed are usually recruited from the streets which means that the criminal organizations can easily tolerate their arrest. The associated losses have already been discounted in their price calculations (Jeschke 1987).

The restructuring of the European cocaine market from peer-oriented markets to organized criminal hierarchies is confirmed by Lewis (1989). He recognized that an established market for cocaine has existed in Europe for twenty years with a considerable elasticity of demand. Over-production in Latin America in recent years is contributing to increased availability and consumption of cocaine in Europe. Moreover, professional criminals are becoming increasingly involved. Lewis advances the interesting conclusion: 'The impact of the illicit global cocaine market ... has been more profound than the effects of the drug itself. The significance of the cocaine market in

Europe lies as much in its effect upon criminal organizations as in any problematic consumption that may occur.' A dramatic event indicating this is the murder of Italian anti-mafia official Judge Giovanni Falcone, in May 1992. This murder seems to have been the consequence, in part, of one of his last investigations involving the case of the transport of 600 kilos of cocaine directly from Columbia to Sicily. The implication of this case is that the Sicilian mafia has been joining forces with the Columbian cartels to further open the pipeline of cocaine to Europe (National Drug Strategy Network 1992). Disclosures since the murder support the theory that this new pipeline involves shipments from Colombia to Europe and then on to the United States and Canada. Operation Green Ice, planned by the Italian judicial authorities, has focused on intercepting this new cocaine supply route (Phillips and Frost 1992, Bone 1992, Van der Putten 1992).

As far as distribution is concerned, on the one hand there is an overlap with the dealership of other kinds of drugs such as cannabis, heroin and amphetamines. This is especially apparent in police intelligence reports which state that cocaine is frequently used as illegal tender (i.e. money equivalent) in transactions involving other drugs, especially heroin. On the other hand, the current impression is that separate networks exist for cocaine too. Some dealers serving particular cocaine circles withhold their services from junkies and cannabis users (Arlacchi 1986, Intraval 1991). The Italian market of illegal drugs has been studied by Arlacchi and Lewis (1989), with methodologies supported by ethnography and by the examination of legal material. They divided the distribution system into two main areas: one competitive and one oligopolistic. The latter is formed by a small number of criminal firms selling drugs to the smaller units of the competitive area, the only ones in contact with the public. In particular, Arlacchi and Lewis define six links in the dealing chain: importers, wholesalers, distributors, pushers selling by weight, street pushers, consumers and consumer-pushers. A likely hypothesis is that the importation, wholesale market and part of the distribution, are structurally the same for both heroin and cocaine. Unlike heroin, however, cocaine is frequently used by suppliers at these levels. Distribution forms at the lowest levels vary according to the type of consumer (Arlacchi 1986).

3.2   Demand

When we compare the demand side with the illegal and criminal supply of cocaine, there are two reasons to regard cocaine as a problem. First, cocaine is a drug in the sense that the substance influences consciousness. This influence may be strong. Excessive use can cause social and health problems. The exact nature of dependency on cocaine is still the subject of discussion. Secondly, its illegal character means that the user must pay a high price for the drug. When this cannot be financed out of legal sources, people may turn to irregular, including illegal, sources. This leads to income-generating crime. Furthermore, the drug can fulfil functions in social settings where rule-breaking behaviour occurs or is even the norm. In avant garde artistic circles, for example, cocaine has always played a role. Cocaine appears, in various ways, to be capable of producing a disinhibiting effect, and to lower the threshold both for criminal behaviour and the use of other drugs, e.g. alcohol and heroin. In the Netherlands Intraval (1989 and 1991) identified cocaine as a so-called entry-drug to heroin addiction. Similar findings have emerged from the United States, from, among others, Petersen et al (1983).

It is common practice for exotic drugs to be surrounded by myths. In the case of cocaine these concern, among other things, its effects on both intellectual performance and sexual enjoyment. Freud, it is suggested, would never have succeeded in formulating his psycho-analytical theory without the help of cocaine (Zuidhof 1984). Earlier, in the second half of the 19th century, the drug was administered to (black) mine workers in the southern United States to increase their productivity. This, incidentally, gave birth to the story that blacks could not control their sexual urges while under the influence of cocaine, which supposedly led them to rape white women. In addition, people thought they were immune to certain types of police bullets (Tieman 1981, Musto 1973, Escohotado 1989).

In recent decades, cocaine was initially regarded as the caviar or champagne among drugs; its status was generally positive. As early as 1953, the American behavioral research scientist Wikler suggested that cocaine use went hand-in-hand with success-oriented behaviour and identity; in other words, it complemented the American culture (Siegel 1984). A quarter of a century later, this view became reality on a large scale in the United States, and later in Europe. The increase in cocaine use in recent years is partly due to the fact that it fits the spirit of the times; the believe that the whole of post-modern society (American and European) has become 'more cocaine-ish'. The drug provides an aura of really having made it combined with the capacity to achieve still more. This is in stark contrast to heroin, the escape route from reality, the symbol of underprivileged youth (Van Ree and Esseveld 1985).

According to several large scale American studies of cocaine users in the Eighties, the high-risk group for cocaine is, unlike heroin, not made up of weak, emotionally disturbed or socially deprived people but strong and resourceful individuals. This sort of person subscribes to the current values of western society; material success, career, ambition, competition. He can be found in the world of advertising and consulting, among actors, writers, lawyers, and the traditional professions. According to these studies, a large portion of heavy cocaine users consists of successful, highly educated yuppies in their twenties and thirties. They have no money worries (Spotts and Shontz 1984, NIDA 1986, Morningstar and Chitwood 1987). In the United States, however, people are beginning to believe that this is, to a large extent, based on myths surrounding cocaine which no longer correspond to reality (Kozel and Adams 1985).

In both America and Europe the image of cocaine has been greatly tarnished in recent years (Arif 1987, Cohen 1989). Nowadays, cocaine is thought to be used in all strata of society and by various categories of people including heroin addicts, middle class youngsters and professionals. Cocaine use is, for example, increasingly seen among opiate addicts (Intraval 1991, SEIT 1988-1991). Furthermore, a third of the non-deviant cocaine users in Amsterdam proved to have no regular job suggesting that cocaine is not always associated with success (Cohen 1989). Recently, more and more young people in the United States, especially those from the ghettos, are becoming addicted to crack cocaine. This has tended to make cocaine associated with a lifestyle of desperation, violence and deprivation.

3.3 Prevalence

The recent trends in cocaine usage in Europe reflect earlier developments in the United States. In many American reports cocaine has been cited as the number one illegal drug. In 1986, The National Institute on Drug Abuse estimated that more than 20 million Americans (nearly 10% of the population) had taken cocaine at least once; that 4-6 million used it regularly; and that between 200,000 and 1 million people would be classified as compulsive users (NIDA 1986). In 1982 as many as 19% of the age group between 18 and 25 years were estimated to have used cocaine at least once in the preceding year. The figures in 1986 for New York (18 years and older) were: 13% had used cocaine at least once; 5% in the last month (Des Jarlais and Friedman 1989). More recent American figures estimate the lifetime prevalence at 11% and the prevalence of current use at 1% (NIDA 1990, Rouse 1991). According to Anthony (1992) the downward trend in the use of cocaine and other illegal drugs observed in the United States is caused by a demographic shift with a decrease in the number of young people and an increase in the number of older people. However, while recent data indicate a trend towards a reduction in recreational use, the number of persons using cocaine at least once a week has risen by 33% between 1985 and 1988 (NIDA 1990). This suggests that cocaine dependence will continue to be a major public health problem in the United States during the 1990s.

Despite the American downward trend, the European prevalence figures are still generally lower than the United States figures. Important national differences seem to exist in Europe. We see, for example, that a relatively large amount of information on the national and regional prevalence of cocaine use has been documented in Spain, far more than in Italy and the Netherlands.

Information on prevalence in the Netherlands is restricted to general household and school samples conducted in Amsterdam, a city hardly representative of the overall Dutch population (see Plomp et al 1990, Sandwijk et al 1988). In Italy, the only national level data that is somewhat related to prevalence are police reports on seizures and studies of clinical populations. The recent study of Ponti et al (1991) at the Department of Forensic Medicine at the University of Milan found that between the years 1986 - 1989 a toxicological examination of 1,248 deaths of accidents, suicides and murder found 3.13% were positive for cocaine. In another study of 1,400 psychiatric reports concerning defendants on trial, 3.53% were found to be cocaine positive. Since these are special populations, the prevalence in the general Italian population can be assumed to be considerably lower. In 1987-88 in Turin, interviews were conducted with 249 young people randomly selected from the general population and 300 treatment clients (opiate addicts). The results indicated that 3.2% of the randomly selected young people used cocaine at least once a week (0.4% more than once a week) compared to 87% of the treatment clients (42.4% more than once a week). From this information, it can be inferred that 1,000 young people and 3,000 heroin addicts in Turin use cocaine more than once a week (Burroni et al 1989). Data from the Turin Health Service for Drug Addicts reveal that 0.2% of the 6,034 individuals registered between 1978 and 1989 primarily took cocaine whereas 3.4% reported they used cocaine combined with heroin. These percentages have not changed significantly over time, though an increase in cocaine use was observed among treatment clients in 1980 and 1981 (Merlo 1990).

Dutch data collected in 1983 show 3% in the age range 15-24 years had used cocaine at least once. The figure for the larger cities was 6%. In the same period, incidentally, a figure of 4% was found for the same age group in West Germany. More recent figures present a stable picture. In 1987, the lifetime prevalence in Amsterdam was 5.8% in the 15-39 years age group (Sandwijk et al 1988). In the wealthy Amsterdam suburb of Het Gooi it was only 2% (Korf et al 1990). The current year prevalence in Amsterdam was 1.7% and the current month prevalence was only 0.6%. Between 1987 and 1990 the percentage of the population of 12 years and over in Amsterdam who had taken cocaine at least once decreased from 5.6% to 5.3%; the current year prevalence decreased from 1.6% to 1.3% (Sandwijk et al 1991). If we look exclusively at acknowledged addiction problems in 1988, more than 900 clients were registered in Dutch CADs (Alcohol and Drug Counselling Centres) in connection with cocaine addiction (NVC 1990). According to the Rotterdam CAD about 90 people using only cocaine sought help for cocaine dependency problems. However, in 1991, the Rotterdam Municipal Health Drugs Services Register (RODIS) records 66% (N=813) of the opiate using clients reporting cocaine use (Toet and Geurs 1992).

Table 3.1 shows the results of different Spanish surveys on the prevalence of cocaine use between 1985 and 1990. In general, the population surveys show a relative stability of prevalence levels since 1985. In Spain in general, and Catalonia in particular, approximately 3% (ranging from 2.9% to 3.7%) have used cocaine at least once in their lifetimes. In Madrid, a city comparable to Barcelona, the lifetime prevalence is 6.1%. Nevertheless, there is evidence that cocaine use is on the increase. In the period 1986 to 1990, lifetime prevalence of the age group 15-64 years increased from 4.7% to 5.7% in Catalonia (see also Barrio et al 1992, PNSD 1992). In all of these surveys, the prevalence of cocaine use is relatively higher than that of heroin. The conclusions of a study of drug use in Madrid assert that the levels of experience and regular use of weekly current cocaine is second in importance only to cannabis (Alvira and Comas 1990).

Table 3.1   Prevalence Spain: 1985-1990 (in %)

Study*	Year	Population age	Lifetime	Last year	Last 6 months	Last month
A	1985	12 and +	3.7	---	1.8	---
B	1986	12 and +	2.4	---	1.2	0.7
C (a)	1986	15-29	4.7	3.2	---	0.2
C (b)	1986	15-64	1.6	1.2	---	0.1
D	1988	18 and +	3.0	---	---	---
E	1989	16 and +	3.0	---	---	---
F	1989	15-64	6.1	---	---	0.9
G (a)	1990	15-29	5.7	3.1	---	1.5
G (b)	1990	15-64	3.3	1.7	---	0.7
G (c)	1990	15 and +	2.9	1.5	---	0.6

*   A EDIS. Context: Spain excluding Ceuta and Melilla. N=6,000.
    B EDIS. Context: Community of Aragon. N=1,800.
    C Departament de Sanitat. Context: Catalonia. a) N=1,200. b) N=1,500.
    D CIS. Context: Spain excluding Ceuta and Melilla. N=2,142.
    E CIS. Context: Spain excluding Ceuta and Melilla. N=2,600.
    F Alvira & Comas. Context: Madrid. N=8,000.
    G Departament de Sanitat. Context: Catalonia. a) N=3,117 b) N=1,560 c) N=1,843.

The figures of admissions for treatment in Spanish health service clinics show an increase in the number of admissions linked to opiate and cocaine use between 1987 and 1991. In Catalonia, the number of persons admitted for treatment increased from 1,513 to 4,489 for heroin and from 41 to 218 for cocaine. In the whole of Spain the number of admissions increased from 10,141 to 30,347 for heroin and from 197 to 989 for cocaine (Organ Tècnic de Drogodependències 1992, PNSD 1992, SIDC 1992). In order to interpret these indicators correctly, one must take into account that the increase may, in part, be due to the increase of the number of treatment centres (PNSD 1992). In 1991, the number of admissions for cocaine continued to represent a relatively small proportion compared to opiates and other drugs. In that year, admissions for cocaine are 3.2% for the whole of Spain, 4.8% in Catalonia and 4.5% in Barcelona (PNSD 1992, SIDC 1992). Furthermore, a progressive increase in the consumption of cocaine among heroin addicts who are admitted to treatment can be observed in Spain. The number of heroin addicts taking cocaine 30 days prior to treatment rose from 42.8% in 1987 to 51.5% in 1991 (PNSD 1992).

Concluding this overview of cocaine prevalence in Europe, it is important to realize that, compared to the United States, the low prevalence of cocaine use and of clinical cocaine dependency in the general population may be a question of time. A relatively long period may elapse from onset of use to the appearance of dependency symptoms in both the person and in society. This may especially be the case if cocaine is used in less potent forms and less dangerous ways. Already there are clear signs of an increase in cocaine in the polydrug population of Europe. We will know if this corresponds to an increase in the general population only by the establishment of epidemiological monitors and the periodic repeat of general surveys and targeted community studies.

3.4   Pharmacology and addiction

Effects of cocaine

Cocaine initially produces profound subjective well-being together with alertness. This is related to relatively specific psychological effects of combining stimulation with ego expansion (Spotts and Shontz 1984). The fundamental effect of cocaine is a magnification of the intensity of almost all normal pleasures. Emotional and sexual feelings are enhanced as well as the feeling of being able to think extraordinarily quickly and clearly. The environment takes on an intensified and lucid quality. In addition, self-confidence and the self-perception of mastery increase. Social inhibitions are reduced and interpersonal communication is facilitated (Van Dyke and Byck 1982, Spotts and Shontz 1984, Gawin and Kleber 1986, Van Limbeek 1986, Van Epen 1988, Johanson and Fishman 1989, Gawin 1991). It must be noted, however, that some individuals who experiment with cocaine describe overwhelming anxiety rather than euphoria as the main effect cocaine had on them (Gawin 1991).

The acute physiological effects of cocaine are hypertension, tachycardia, vasoconstriction, mydriasis and a rise in body temperature. Tremors and convulsions may occur. It is also known that muscular strength and endurance can temporally increase together with a reduction in appetite (Van Epen 1988, Volpe 1992). The immediate effect of cocaine is very rewarding. The emotional state of the user can swing rapidly from one extreme to the other. According to Peterson et al (1983), a direct effect is to be expected for only half an hour. The effect of taking cocaine intranasally differs, however, from other routes. The effect starts gradually and lasts for more than a half hour. If the purity of the cocaine is high, the effect can sometimes be noticed hours after it has been taken. When cocaine is injected intravenously or taken by way of free basing its effect starts almost immediately but is short-lasting. The effect of cocaine can pass within minutes. This is especially true when cocaine is taken by way of free basing. According to this method, cocaine is smoked in a glass pipe sometimes filled with an alcoholic drink (usually strong rum). It also can be done with the aid of a (drinking) glass covered with foil (Van Epen 1988).

Forms of cocaine preparations

Cocaine is an alkaloid (benzoylmethylecgonine) derived from the leaves of the Erythroxylon coca (Farrar and Kearns 1989). Cocaine is available in two forms: cocaine hydrochloride and highly purified cocaine alkaloid (free base). The first form, cocaine hydrochloride, is generally taken intranasally or intravenously. It is heat labile but water soluble. The second form is derived from cocaine hydrochloride. When sodium bicarbonate is added to a watery solution of cocaine hydrochloride and heated, cocaine alkaloid is liberated. It can be picked up as hard paste, once the water is evaporated, or as free base cocaine through extraction by means of ether. Because it is heat stable and insoluble in water, cocaine alkaloid is generally taken by vapour inhalation (smoking). When it is smoked, the crystals break into small pieces making a popping sound. The onomatopoeic term crack is derived from this. (Volpe 1992). Compared to the United States, crack is not found on a great scale on the streets of Europe.

Psychopharmacology

The reinforcing effect of a drug is the effect that increases the probability that the drug will be self-administered again. Cocaine has been found to be a highly reinforcing drug in both animals and humans (Schuster et al 1981, Wise 1984). In various animal-studies, cocaine functions as a more powerful reinforcer than, for example, heroin. In rats, the fatality incidence associated with continuous cocaine self-administration is 90%, in contrast to 36% with heroin. In self-administration experiments, monkeys will do more work to receive cocaine than other drugs and choose cocaine over food and even over the opportunity to socialize with other monkeys (Johanson and Fishman 1989). The immediate effects of cocaine are highly rewarding. When asked the difference, opiate addicts often say that they really need heroin and/or methadone but not cocaine. Nevertheless, once the compulsive user starts taking cocaine he keeps on seeking it. According to Van Meerten (1992), cocaine appears to give the user a brief moment of complete satisfaction followed by a rock bottom that calls for an immediate repeat. The user wants it again and again. In this sense cocaine is a seductive and intensely compelling drug. Spotts and Shontz name it the most unsatisfying and limitless stimulant in existence (1984).

Neuropharmacology

Cocaine has pronounced effects on both the peripheral and central nervous systems. In summarizing the literature on this subject, Volpe (1992) describes four neurochemical mechanisms of action (see also Knapp and Mandell 1972, Van Dyke and Byck 1982, Gawin and Ellinwood 1988, Farrar and Kearns 1989, Kuhar et al 1991).

a. Cocaine impairs the reuptake of epinephrine and norepinephrine by presynaptic nerve endings, causing a significant activation of adrenergic systems. This activation causes the acute physical effects of cocaine, e.g. hypertension, tachycardia and vasoconstriction.
b. Cocaine impairs the reuptake of dopamine. This causes the activation of dopaminergic systems crucial to the sense of euphoria that follows cocaine intake. It involves multiple brain pathways. With longer term use, however, dopamine becomes progressively depleted from the nerve endings which may lead to a sense of dysphoria. This symptom is prominent during withdrawal and relates to the subsequent craving characteristic for the drug. There is a growing evidence that the effects of cocaine on dopaminergic neurons are closely associated with its rein-forcing properties. While this dopamine hypotheses is strongly supported in animal studies of drug self-administration, the extent of its involve-ment in human cocaine dependence has not been fully elucidated (Fibi-ger et al 1992, Kuhar 1992).
c. By blocking the uptake of tryptophan and serotonin, cocaine impairs the homeostasis of the serotonic system which may account for the striking alteration in the sleep-waking cycle and may also enhance the general excitatory effects of dopamine.
d. Cocaine interrupts the initiation and propagation of peripheral nerve impulses by preventing the increase in permeability of sodium ions. This is the effect that underlies the local anaesthetic action of cocaine. As yet it is not clear whether this disturbance in ion-permeability also causes some of the cerebrovascular effects of cocaine.

Addiction

For several decades drug dependency has largely been equated with an avoidance of physical discomfort on withdrawal. This is reflected in the classic concepts in drug and alcohol addiction theory, such as physiological with-drawal, tolerance, and physical dependence. Cocaine, however, does not produce gross physiological withdrawal symptoms. By and large, it is also thought that tolerance scarcely ever develops with cocaine. The user may want increasing amounts during a given run, but this apparently has no influence on the amount needed the next time to satisfy his needs. Nevertheless, drug-seeking behaviour does manifest itself in cocaine users. More important, their craving for the drug can be enormous. This has led to the belief that cocaine addiction is psychological rather than physiological. For many years, both cocaine users and scientists viewed cocaine addiction as an intense habituation to the very pleasant euphoriant effects of the drug. Recent studies makes it increasingly clear, however, that chronic drug use leads to neurophysiological adaptation. With cocaine, this does not lead to the classic abstinence syndrome that can be observed in opiate addiction. According to Gawin (1991), 'chronic high dose use of cocaine may generate sustained neurological changes in the brain systems that regulate only psychological processes in particular hedonic responsivity or pleasure'. These changes produce a physiological addiction and withdrawal syndrome whose clinical expression may be solely psychological (Gawin and Kleber 1986, Gawin and Ellinwood 1989, Gawin 1991).

Post-cocaine syndrome

Gawin and Kleber (1986) were among the first contemporary clinicians to describe a syndrome that was specific to problematic cocaine use. They identified a sequence of post-cocaine phases that characterized the psycho-physiological pathology that were present in patients reporting cocaine abuse. In a later clarification, Gawin (1991) describes a triphasic pattern of symptoms indicative of the unique response to abstinence from cocaine (see also Siegel 1985, Gawin and Ellinwood 1988, Waldorf et al 1991). Gawin referred to the syndrome as abstinence phases, but we retained the construction of post-cocaine phases because it is more descriptive and dynamic than the term absti-nence would suggest. Thus, many of these symptoms may appear without indicating abstinence from cocaine as when, for example, the patterns of use change. Diagram 3.1 shows Gawin's latest published diagram. The three phases in diagram 3.1 can be explained as follows.

1. Crash

Immediately after a cocaine binge there is a crash of mood and energy which parallels an alcohol hangover. Cocaine craving and depression are characteristic for this phase. Agitation and anxiety rapidly intensify and suspiciousness and paranoia may be prominent. During the next one to four hours, cocaine craving decreases, while exhaustion and craving for sleep increases.

2. Withdrawal

Shortly after the crash, a feeling of dysphoria increases. The dysphoric syn-drome is characterized by decreased activity, anxiety, lack of motivation and boredom, with markedly diminished intensity of normal pleasurable experiences (anhedonia). When compared to memories of cocaine-induced euphoria, this delimited state induces severe cocaine cravings. Except for the fact that there are no (gross) physiological changes, this withdrawal phase parallels withdrawal from other substances. Cocaine users wanting to stop are usually able to withstand this anhedonic dysphoria, until they are presented with a conditioned cue. This cue superimposes a second dimension (evoked or conditioned craving) on anhedonic craving and cocaine use often resumes. Conditioned cravings start after the appearance of varied objects or events, such as specific persons, specific locations, specific moods, or 'use' objects such a razor blades, money, and white powder. The appearance of these objects or events are experienced as partial memories of cocaine euphoria. Since cocaine is found to be such a powerful reinforcer in both animals and humans, it is not surprising that conditioned craving is more intense in cocaine dependence than in any other addiction.

3. Extinction

According to Gawin, intermittent, conditioned cocaine craving can still emerge months or even years after the last cocaine use. In this sense (lasting) cocaine abstinence can be seen as experiencing conditioned cravings without relapse. Craving is gradually then extinguished: the consistent pairing of cues with cocaine euphoria does not occur.

Diagram 3.1 Post-cocaine phases (adopted from Gawin 1991)

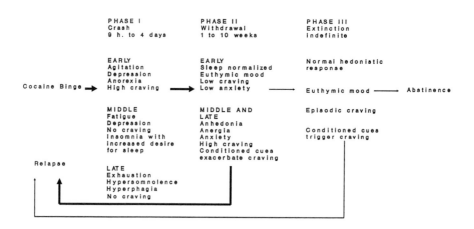

3.5   Morbidity and mortality

The statistics on the morbidity and mortality in Europe due to cocaine use are scarce. Much of the data relating to cocaine is obscured by factors such as poly-drug use and unclear, inconsistent monitoring of the cocaine medical situation. This is, in part, due to the idea that cocaine is not yet a significant public health problem in Europe and therefore does not require separate mention. Thus, for example, while in the Netherlands there are no consistent drug death or emergency admission monitoring systems, limited systems do exist Spain and Italy. The sum of these limitations makes the description and comparison of morbidity and mortality data for cocaine very difficult.

The explosion of biomedical research on cocaine, especially in the United States, has resulted in the detection of new morbid reactions to cocaine use. The psychiatric morbidity associated with cocaine has been well diagnosed for twenty years: depression, hyperarousal, hallucinations and paranoia are representative symptoms (Post 1975). Benowitz (1992) has provided a useful overview of the medical complications of cocaine intoxication and abuse. Several of these have caused intense public concern in the United States. Thus, one of the reproductive disturbances and neonatal effects occurring in some infants born of cocaine-abusing mothers is the so-called crack baby syndrome. The symptoms include neurobehavioural abnormalities including abnormal sleep-wake cycles, tremulousness, poor feeding and hyperreflexia (Oro and Dixon 1987). Cardiovascular complications have also received a great deal of American attention due to the sudden death of celebrities such as sports stars. Cocaine induced myocardial infarction has even become commonplace in American hospital emergency treatment rooms (Isner and Chokshi 1991). Central nervous system, respiratory, infectious and metabolic morbidities have all been documented in medical literature.

In the United States, cocaine is associated with more emergency department visits and more deaths than any other drug of abuse. Most of the cocaine deaths recorded in the medical literature are of Americans. These deaths tend to be sudden. A review of the American cases suggest that these deaths may be due more to trauma than to acute intoxication (Tardiff et al 1989). It is interesting to note that a breakdown of these cocaine-related deaths show that the top causes were homicide (37.5%) and acute narcotism (combination of heroin and cocaine; 12%). Only 11.6% of these deaths were directly related to cocaine intoxication. The absence of conditions inducing trauma may account for the virtually unknown prevalence of cocaine deaths in Europe. Also, given the relatively high prevalence of acute narcotism cocaine-related deaths, many European cases may have been recorded as heroin overdose cases.

Since 1987, the Turin Central Coordinating Office for Drug Dependence of the Municipal Health Unit has been routinely collecting data on overdose-related emergency room episodes occurring in the five city hospitals. Furthermore, since 1978, the unit has been compiling data on deaths from opiate overdoses. In 1991, 1,339 emergency room episodes and 75 deaths from opiate overdoses were recorded. No episodes or deaths connected with cocaine use had been recorded either in 1991 or the preceding years. These data require several qualifying remarks. There is no routine test for psychoactive substances contained in the body fluids. The result is that the cocaine (and other drug) mortality and morbidity are likely to be highly underestimated in emergency rooms. Unlike heroin, there is usually no useful clue to support a hypothesis of cocaine abuse.

As we said earlier, there is no consequent drug death or hospital emergency room monitoring system in the Netherlands. In 1991, 9 overdose deaths were registered in Rotterdam. In stating the cause of death, however, no distinction was made between cocaine or other kinds of drugs. Most of the deaths were related to opiates. The role played by cocaine is not clear. Between 1984 and 1986, Amsterdam listed 175 cases of drug-deaths. Cocaine, to some extent in combination with alcohol, was identified in three cases (Van Brussel 1991).

In Spain, (where a hospital emergency room monitoring system does exist), the period 1987 to 1990 saw a rise in the number of hospital emergencies brought about, directly or indirectly, by use of cocaine, with or without other drugs. This increase indicates that cocaine is gaining in relative importance in the emergency room monitoring system; i.e. from 1.0% of the total cases in 1987 to 3.3% in 1990 (SEIT 1988-1991). A study carried out in three Spanish cities (Madrid, Barcelona and Valencia) between 1983 and 1989 investigated the change in mortality due to an acute response resulting from drug use Rodríguez (1989). This research showed that mortality rates have tended to increase since 1986. Cocaine was detected in 17% of blood samples and in 26% of urine samples. However, when we look at the American acute narcotism cases, in only 0.7% of the blood samples was cocaine the only drug detected (Barrio et al 1990). In general, most of the deaths (more than 90%) were related to heroin (PNSD 1992). In Barcelona, in a study of blood samples obtained from 124 deceased people in 1990, 76 samples contained heroin (61%) and 22 (29%) cocaine. Of the 76 samples containing heroin, 15 (20%) also contained cocaine (SIDB 1991).

3.6 Social risks and criminality

Arnao (1991) has suggested that when considering the health and social risks of cocaine, two crucial questions must be asked: what are the levels and modalities of cocaine use that lead to social risks derived from the cultural context and what are the risks of the onset of physical and mental problems. Arnao concludes that in terms of a risk analysis based on data from America and the Netherlands, cocaine can be properly compared to alcohol and alcoholism. However, cocaine, at least in the United States, has been associated with crime and violence while this aspect has been played down in relation to alcohol. Goldstein et al (1989) investigated the relationship between crack cocaine and homicide in New York specifying exactly how many homicides were related to the pharmacological effects of crack cocaine and how many to social causes. The research supported Arnao's theory. Alcohol was more often associated with homicides attributable to pharmacological effects than crack cocaine. The risks of homicide related to cocaine were associated with market competition factors. The involvement in the system of crack supply was the principle cause of cocaine-related homicide.

The New York results are also supported in a recent study of cocaine and criminality in Milan. Ponti et al (1991) found that while a connection of ambience could be identified between cocaine and criminality, this relationship was not be found to have either a direct or indirect correlation. In strong contrast to heroin, there is no symbiotic relationship between cocaine and crime. Rather, criminals use cocaine for hedonistic and performance-enhancing reasons in ways similar to non-criminal cocaine users. The New York and Milan studies both suggest that the social risks of cocaine will increase not primarily from the behaviour of the cocaine users themselves, but from changes in the market and supply of cocaine that implicate the user.

However, counter-indicators can also be found in the literature. Some authors state that physical offenses against persons are more common with cocaine than with heroin. Other criminal consequences have mainly to do with the uninhibiting effect of cocaine: macho and daredevil behaviour, and physical and psychological self-overestimation (Spotts and Shontz 1980, McBride 1981). In the matter of cocaine use and violent crime, it is generally thought that alcohol, benzodiazepines, and substances that depress, rather than stimulate consciousness play a greater role than cocaine in violent crime. Cocaine in combination with other substances may, however, function as a potentiator in this context. Analysis of incidental cases has shown that cocaine can play a substantial role in violent crime, again in combination with alcohol (Van Epen 1988, Budd 1989).

3.7   Sociocultural studies

Studies of clinical populations are not, as such, a useful means of obtaining insight into problematic use patterns that do not present a clinical dependency syndrome. Furthermore, these studies tell little of the sociocultural contexts of use, the functions and meaning that cocaine has for the user. Sociocultural studies have emerged to fill this gap in knowledge and identify a wider range of patterns and consequences relevant to prevention and intervention. Several important community-based field surveys have been published (Chitwood

1985, Cohen 1989, Erickson and Murray 1989, Macchia et al 1990, Van Meter 1990). These community-based surveys of cocaine use provide help in defining subtle patterns and furnish evidence supporting the general population survey results that cocaine dependency, like clinical alcoholism, is an exceptional case. An Amsterdam community-based survey of cocaine users from a non-deviant population (i.e. no recent prison or drug treatment history) employed an index of level of cocaine use first developed in Miami (Morningstar and Chitwood 1983 and 1984, Chitwood 1985, Cohen 1989). This index gives equal weight to three factors: self-reported route of ingestion; frequency of use; and quantity of use, and produces a low-medium-high measure of level of use. Taken in the perspective of restricted use over time, this measure profiled six distinct developmental patterns of cocaine use: 1. first much cocaine-slowly less; 2. slowly more cocaine; 3. stable level from first to current use; 4. up-top-down, i.e. first a relatively little cocaine use that rises to a peak and then drops to first levels; 5. intermittent use at the same level, but with significant breaks in time; 6. varying cocaine use with peaks and troughs in an irregular trajectory. In the Amsterdam study, most prevalent were Pattern 4 (39%) and Pattern 6 (33%). Only 3% of the sample presented Pattern 2, the pattern characteristic of opiate dependence (Cohen 1989). The earlier Miami study reported that approximately half (48%) of the sample were users at a level higher than the first year, one-fifth remained low-level users and nearly as many (18%) progressed from low-level use during their first year to higher levels in subsequent years (Chitwood 1985). The Miami results are comparable to the Amsterdam ones: a high prevalence of patterns characterized by variation in use level. Another similarity to the Amsterdam study was that only a relatively small percentage of the sample progressed in a way characteristic of opiate dependency; from initial low to current high level of use (12%).

In addition to these community-based surveys, an eleven year follow-up of a friendship network of 27 San Franciscan cocaine users has been published (Murphy et al 1989). This unique study presents complementary data concerning long-term patterns of cocaine use. Murphy and co-workers use the term casual users to refer to their subjects first interviewed in 1974-1975. The vast majority of these subjects did not use cocaine daily nor did they believe cocaine to be addictive. They had experienced only a few negative effects from using the drug. At the follow-up, eleven years later, Murphy and co-workers found four distinct patterns of variation over time (and their frequency in the sample): continuous controlled use (33%); progression from controlled to heavy to controlled use (33%); from controlled to heavy to abstinence (24%); and from controlled use to abstinence (10%). They also report on the exception, one person who continued heavy use. These findings can be contrasted with heroin where continued heavy use is perhaps the most prevalent pattern and continuous controlled use the exception (see Zinberg 1984).

In adopting the terms controlled and heavy use, Murphy and co-workers articulated the value of an ethnographic approach for sociocultural studies. In ethnography, the subculture of the cocaine users themselves are primary (Adler and Adler 1987). The San Francisco group has recently published an extensive ethnography of 228 heavy cocaine users describing and analyzing a myriad of subtle patterns and variations of cocaine use that underlie the changes that cocaine users themselves experience (Waldorf et al 1991). They postulate two integrating themes that, taken together, explain why so many heavy users can take cocaine without incurring problems and why, when they do have problems, they are able to surmount them. The themes include having a stake in conventional life and identity, and the transformation of the 'high' in such a way that the pleasure of taking cocaine diminishes. This work deepens and elaborates on the simple four-fold classification of cocaine users into: experimenters, social-recreational, involved, and dysfunctional, which has been the mainstay in the literature for the last two decades (see Inciardi 1986).

These ethnographic sociocultural studies are being added to by a growing number of field studies in America analyzing the factors and contexts of crack cocaine use. Many of these ethnographics place the crack cocaine phenomenon in the struggles and opportunities of American ghetto and underclass life and identify the meaning of crack cocaine to its young user-dealers as a modern heroine myth (Bourgois 1989, Williams 1989). Crack cocaine in these studies becomes a symbol of struggle between the underclasses of American society and an uncaring social system which offers little alternatives for poor youth.

In Europe, to date, there are few published studies that analyze the sociocultural factors of cocaine use in any depth. The work that has been done has focused mainly upon the cocaine use patterns and rituals of heroin addicts (Grund et al 1991). While these studies are useful in documenting the devastating effect cocaine has had on the heroin addict population, the results cannot be generalized to other populations where cocaine is also frequently used. There is a need in Europe for real research which complements the small, but growing, number of quantitative analyses of the sociocultural patterns of use, with rigorous qualitative analyses of cocaine use in different circuits in the general population. This book is, in part, aimed at filling the gap.

3.8 Socio-historical studies

The history of cocaine in Europe is as old as the drug itself. Coca leaves originate mainly from fields in Bolivia, Peru and Colombia. The indigenous Indian population had discovered the stimulating effect for centuries and coca had a traditional place in their religious and medical practices. Coca leaves were chewed while working and as a remedy against fatigue, pain and appetite. Later, the Spanish conquerors in South America administered the drug to enslaved Indian mine workers. The active constituent in cocaine was first extracted from the coca leaf in 1859. Subsequently, cocaine was marketed as a remedy for colds, influenza and as a local anaesthetic. By the 1880s, cocaine had firmly entered the commercial market of a wide-range of pharmaceutical and patent medicines, remedies and tonics (Musto 1992). One of the most popular was a French coca extract marketed as a wine under the trade name Vin Mariani. Pope Leo XII was an avid consumer of the wine and awarded Angelo Mariani a gold medal. American products, such as Coca-Cola, joined the market.

In the Netherlands, the 'Nederlandsche Cocaïnefabriek' (Dutch Cocaine Factory) began producing remedies based on cocaine in 1870 (Korf and De Kort 1989). Dutch agricultural scientists became involved in the production of the raw coca leaf. Plantations of coca were developed in the Dutch colony of Indonesia. These produced a leaf twice as potent as that found in the Andes. The German giant pharmaceutical company Merck, of Darmstadt, distributed the drug widely to pharmacies and physicians before the First World War. The drug was often controlled by laws such as the British Pharmacy Act of 1868. Physicians were among the first to become addicted to the cocaine they prescribed.

The American drug historian David Musto (1992) identifies a historical cycle involving cocaine over the last century that fits, with certain modifications, both the European and American historical experience. Cocaine first entered the market with enthusiastic reception as shown in the Vin Mariani case. Later, the public image changed as cocaine abuse became more visible and new celebrity and criminal underworld social groups started using it. A strong negative public opinion emerged and cocaine use died out. This seems to have been the experience of both America and Europe in the period between 1880 and 1930. By 1930, cocaine had been perceived as a significant social and medical problem in the United States, Great Britain, France and Germany and prohibitionist laws were evident on both sides of the Atlantic Ocean.

Today the coca leaf is illegally processed into cocaine primarily in Colombia from where it is brought onto the international market. The Medellin Cartel is the most notorious criminal organisation responsible for marketing (Eddy 1989). The Colombian government receives financial and military support from the United States to assist its fight against the drug organizations. The American market has become saturated in recent years, which is making the market in Western Europe more attractive for international traffickers. Both Europe and America seem to be experiencing the second cocaine epidemic cycle of this century. As Musto points out, in the 1960s when cocaine gradually reappeared on the (illegal) market, it was generally and professionally perceived as a safe and beneficial drug. Thirty years later, America and, at different times, certain European countries, are involved in a war on cocaine trafficking and use. We seem to be once again experiencing the historical cycle of reception and rejection of cocaine.

In order to verify such global theories, there needs to be basic socio-historical research specifically related to cocaine. In Europe, papers appear fragmentally as do chapters in books on topics relevant to the European history of cocaine (e.g. Arnao 1983, Helfand 1988, Escohotado 1989, Korf and De Kort 1989). The sociocultural studies of cocaine indicate a high level of variation and change in cocaine use and users. These variations also appear over relatively long historical spans of time that are greater than the career of the individual cocaine user. Therefore, socio-historical studies are important components of any attempt at understanding cocaine use. This book will also show how cocaine is part of this particular historical period at the end of the Twentieth Century. Historical knowledge is indeed a significant part of any comprehensive analysis of the nature and extent of cocaine or any, for that matter, drug.

CHAPTER 4

METHODOLOGY

This chapter describes the way the studies in the three cities have been carried out. It is important to note that the phenomenon we are dealing with, illegal drug use, is rare. It is a phenomenon affecting a few people in the total population and is, where possible, kept hidden (Spreen 1992a). 'Cocaine users are considered a hidden population, because sampling frames are lacking, and/ or because members are difficult to locate and/or difficult to get in touch with, and/or because it may be hard to determine whether a given individual belongs to the population. The variable that defines the members' use of cocaine is assumed to have a social meaning and, therefore, the existence of a pattern of contact between cocaine users is also assumed to exist' (Spreen 1992b). Another problematic issue in research among cocaine users is that most individuals are reluctant to disclose information, pertaining to the variable(s), that is essential for the definition of the population. Thus, because of the complexity of the phenomenon the use of several different qualitative and quantitative research methods is required. After a short review of the most important methods and techniques usually applied in research on the nature and extent of drug use, special attention will be paid to snowball sampling and network analysis since we consider these to be especially important and novel methods in the field of drug research. In conclusion, the research design of the cocaine studies in the three cities will be explained.

4.1   Methods and techniques in drug research

Certain indicators are frequently used in the field of drug research. Usually, these come from (drug) care agencies, police and courts. There are, for example, the criteria drawn up by the Pompidou Group (1986) which are intended for making comparisons between different European countries. There are also certain monitoring systems. Each of the three cities has its own drug

---

information system: in Barcelona SIDB (Sistema d'Informacio de Drogodependencies de Barcelona), in Rotterdam RODIS (Rotterdam Drug Information System) and in Turin SIT (System Information Turin). These systems enable researchers to monitor drug use by providing information in a constant and updated form. They do not however provide direct estimates of prevalence. Furthermore, most data they gather relates to heroin users (poly-drug users), and there is only little information on small numbers of cocaine users. The same holds true for police and court records. This means that it is impossible to use these information systems and records for estimating the number of cocaine users by applying the nomination technique or the capture-recapture method in the same way as for opiate addicts (Hartnoll et al 1985a and 1985b, Korf and Van Poppel 1986, Intraval 1989 and 1991).

Nomination technique

The nomination technique is an estimation method based on information which individuals in a sample provide about their network of acquaintances. Users are asked to name friends who take hard drugs, and whether these friends have been in touch with drug programmes, health services or any other similar body, in the preceding twelve months. On the basis of the ratio of programme to non-programme and figures from the drug assistance agencies, an estimate can be made of the number of drug users. There was already evidence that neither drug programmes and health service agencies nor police and court records could be used as a frame of reference for people who only use cocaine. In many cases there are no figures at all and any information which is available relates to only a small number of cocaine users. Furthermore, this technique has to be used with caution because it may be the source of later errors. In the first place, we have to make sure that the initial sample on which the nomination is based is truly representative. In the second place, there are problems due to people withholding information or providing misinformation about their acquaintances.

Capture-recapture method

This method, first used in biology studies, consists of comparing two independent samples from the same population. The overlap, the number of people appearing in both samples, is noted. On the basis of this overlap and the size of both samples it is possible to estimate the size of the whole population. If the overlap is considerable, the population is relatively small. If, on the other hand, the overlap is slight, the population is relatively large. In practice, however, especially for inaccessible groups such as drug users, it is extremely difficult to draw two independent samples from the population. A common solution is to use two separate existing data files. For opiate users this generally means files from the police, drug programmes or health services.

However, this method has the same problem as the nomination technique: drug assistance agencies and the police have too few figures available on cocaine users.

Surveys

Representative, or random, samples from general population surveys are used in many areas of research. A section of the population is selected at random. By using probability theory, the information derived from the sample can be generalized to the population as a whole. A necessary condition in estimating extent with this procedure is that the variable to be measured (for instance cocaine use) occurs sufficient frequency in the sample. This will probably not be the case for a rare and hidden population such as cocaine users. There are other specific problems associated with the attempt of estimating the extent of cocaine use from general population surveys, e.g. non-response and difficulties with the location and withholding of information (Hughes et al 1982). Much depends on the probability that a user will admit that he takes cocaine. Possession, transportation and manufacture of cocaine is a criminal offence. Cocaine users risk being arrested for possession of an illegal drug. Thus, the user may be reluctant to admit taking cocaine. This means that the survey method may well result in an underestimation of the true prevalence of use in the population. On the other hand, some people still regard cocaine as a status drug. This means that certain users may tend to exaggerate their true patterns and meanings. Furthermore, since cocaine has, as yet, not such a bad name as other drugs, non-users may say they use cocaine to increase their feelings of self-importance. Given these considerations, to obtain a sufficient number of respondents and an acceptable level of precision, a large and costly sample would have to be drawn. To add to this disadvantage, surveys usually use questions with pre-coded answering categories making the qualitative analysis of in-depth interviews almost impossible. All in all, it can be concluded that a general population survey is not in itself an appropriate method to study the nature and size of cocaine use.

4.2   Network analysis

Social network analysis is an approach to social phenomena in which the behaviour of people (or other units of analysis such as a firm or a country) is regarded as being influenced by their social environment, and vice versa. From the perspective of social network analysis, the social environment can be expressed as patterns or regularities in relationships among interacting units. Many concepts have been developed for the systematic study of the structure of relationships within a given population, like centrality, density, components etc. (for introductory handbooks we refer to Knoke and Kuklinski 1982, Scott 1991, and Wasserman and Faust, in press). A study of this kind, while of undeniable interest, is, in general terms, much more complex than a study based simply on attributes or characteristics. Consider that in a population of N individuals where only one type of binary relationship is studied, this relationship can appear in N(N-1)/2 different cases. For instance, in a small population consisting of 500 individuals 122,500 possible relationships may be observed.

Network analysis is a useful approach for research into cocaine use precisely because cocaine is considered a social drug. This implies that cocaine users will know if their acquaintances are also taking cocaine. Consequently, it will be useful to study aspects of the network structure. Fundamental questions can be formulated such as the degree to which users are aware of each other's cocaine use, the extent to which the user groups are divided into sub-circuits, and the type of people who occupy the central position in such circuits. If the link between two circuits[1] of users is formed by a small number of people, it is interesting to know what sort of people they are. Network information can also shed much light on distribution patterns of cocaine use. For example: are a few individuals with a special role in the world of users responsible for this distribution, or do all users share this distribution task more or less equally? However, most measurement techniques elaborated in social network methodology have been developed for situations in which the entire network is known, or assumed to be known. In the case of research into cocaine use, it is obvious that the entire user population cannot be interviewed or located, so a sample is needed to obtain information on the network.

Network-based sampling methods (link-tracing methods) are frequently used as a tool for locating a reasonable number of respondents in inaccessible populations or in cases of unsanctioned behaviour. However, the unit of analysis has usually been the individual (Sudman et al 1988, Laumann et al 1989, Spreen 1992a). In our research we used a network-based sample for two purposes: locating a reasonable amount of respondents, and obtaining information on certain important aspects of the networks of cocaine users. Our objective was not to make an analysis of the total network, as has frequently been done in the social sciences (intensive analyses of all relationships existing in relatively well-defined worlds; see Mayer 1966, Wheeldon 1969, Bott 1971 Barnes 1989) but rather to construct, by a link-tracing approach, a sample, with a minimum of bias, for obtaining information on relationships between users, which could subsequently be used for estimating the size of the cocaine user populations (Del Castillo 1991, Frank and Snijders 1992). This is why we chose the snowball sampling method in which the respondents as well as the nominees were selected as far as possible at random.

The chains were kept short, although not in Barcelona. According to Snijders (1992) the possible analyses of chains with two (i.e. level one) or more links are always difficult. In Barcelona the chains were traced in greater length:

ten chains went beyond level one (one of them reaching level six). In Rotterdam most of the chains did not reach beyond level one. Techniques were developed for analyzing the relationships between the respondents and nominees, and the role played by cocaine. The focus was on the composition of the personal networks of the respondents, and on the analysis (by a method related to multiple linear regression) of characteristics of relationships, such as the significance of cocaine in the relationship. These techniques provided a means of capturing important aspects of network structure. In Turin the chain linking method could not be applied due to the difficulties caused by the penal framework. For a detailed discussion on sampling and analyzing structure in hidden populations, see appendix C.

4.3   Snowball sampling

The method of snowball sampling was originally developed for analyzing social structures within society. In 1958 Coleman introduced the snowball sample as a method of data collection which also takes account of the social environment. According to Coleman, snowball sampling (including other sampling methods in the family of link-tracing designs) is intended for studying social groups or behaviour in society, with reference to the social structure. He defined the new type of data as sociometric-type data, i.e. data pertaining to the respondent's relationships with specific individuals. To date, this method has been used in drug research mainly for contacting respondents. However, this method can also be used to gain insight into relationship structures in which cocaine plays a role.

Erickson (1979) describes a snowball sample as a chain starting from a first sample in which individuals are asked to name some of their acquaintances, who will constitute the second wave of interviewees. The same questions are put to the second wave in order to construct the third wave, and so on. The most frequently cited references relating to snowball sampling are Goodman (1961) and Biernacki and Waldorf (1981). The first provides a statistical theoretical basis of this method enabling the estimation of certain parameters of social structure studied. Goodman's work presupposes some restrictions, for example that the initial sample (which he calls phase zero) is a random sample of individuals of a population. There are also required constant numbers, e.g. k, of the individuals named by each respondent. This number must be kept constant in each sampling phase. S, the number of phases implemented, is also a required term. In contrast, Biernacki and Waldorf (1981) explain through qualitative field descriptions of the conduct of efficient snowball sampling. Their article is based on an exploratory study of former heroin addicts. This study was essentially qualitative and therefore, they did not feel bound to the statistical requirements· of a random sample or the constant of the number of

---

nominees. In subsequent cocaine studies, such as those done by Kaplan et al (1987), Avico et al (1988) and Cohen (1989), only one or two persons are selected at random from the total number of persons named by each respondent (nominee). Another difference between these cocaine snowball studies and Goodman's is that they extend the chains indefinitely until they die a natural death due to the inability to make new contacts.

The main advantage of this sampling method is that it enables the researcher to obtain samples of a substantial size (between 100 and 200 people) in rare and hidden populations. This enhances the standard of qualitative studies and, at the same time, facilitates the investigation of the social relationship networks of the respondent. The snowball sampling method may also provide the means of estimating the total number of users (Del Castillo 1991, Snijders 1991 and 1992, Frank and Snijders 1992). The starting point is the initial sample of users, each of whom is asked to name all other users known to him. The newly nominated users form the following 'wave' of the snowball and so it goes on. To ensure a statistically valid estimate, the following considerations are important.

- The initial group of users of the snowball sample must resemble as closely as possible a random or stratified random sample of users. If the beginning of the snowball is not strictly random, there is the risk that the results will be biased. It is necessary to take account of the differences in probability that the various groups will be represented in the initial sample.
- In estimating the number of users by the snowball method, the capture-recapture method must be used: how many of the users named in successive waves were already in the sample, and how many are genuine new names?
- The size of the initial group is of great importance for the accuracy of the final estimate. If the initial group is too small, meaningful results will be difficult to obtain. The required size of the initial group will depend, among other things, on the network structure, that is, the pattern of mutual acquaintance among users, and, of course, on the willingness of users to name other users[2]. According to Snijders (1992) 'the initial sample size n of a one-wave snowball sample should not be much smaller than the square root of the population size, in order to make precise statistical inferences from snowball samples'.

Another important aspect is that besides providing an estimate of the extent of use, the snowball method, in the same way as other sampling methods, provides information about the composition of the user population. It should be noted that the snowball sample, in itself, is not a representative group of users: there is a much greater chance of better-known users being included in the sample than less-known users. The researcher must take this bias into account when he interprets the data on the composition of the snowball sample. For a literature list on the use of snowball sampling and related 'ascending' methodologies, see the article of Spreen (1992a).

4.4   Qualitative methodology and ethnographic field work

Qualitative methodology, based on the phenomenological tradition of the social sciences, seeks to obtain an understanding of social phenomena from the actor's perspective. It has been frequently used, from many different theoretical perspectives in sociology or anthropology, for in-depth studies of so-called primitive communities, marginal groups, criminal and outlaw subcultures (see Goffman 1961, Becker 1963, Wax 1971, Schatzman et al 1973, Fiedler 1978, Schein 1987, Werner 1988). Qualitative research is often associated with intensive fieldwork involving a first hand approach to reality and a thick description using ethnographic techniques. Fieldwork involves strategies for entering into and maintaining research relationships, with emphasis on the effect of interactions in the field on the perspective of the researcher.

The use of qualitative techniques has been developed mainly by anthropologists. A very important part of such a technique is anthropological fieldwork in which the researcher has direct contact for a certain period of time with the reality he is studying. From his position as outside observer, the researcher has to gather information on the vision the social actors have of themselves and compare this objectively with existing theories and different sources of information. In the ideal case, reality is understood through the social actors' categories, definitions and values (Malinowski 1967, Barley 1983). Díaz et al (1992) discuss ethnographic field work in an urban context. In the qualitative approach, a variety of research techniques is normally used: direct and participant observation, open and semi-open interviews, focused life stories, discussion groups, analysis of social networks, etc. The selection of one technique rather than another and the degree to which these techniques overlap depend on the purpose of the research question, the limits of the project and the nature of the subject being studied. Examples of the use of qualitative techniques in the field of drugs research are found in Medina-Mora et al 1980, Funes and Romaní 1985, Korf and Van Poppel 1986, Romaní et al 1989 and 1991, Intraval 1989 and 1991, Swierstra 1990, Grapendaal et al 1991.

Qualitative methodology is the most suitable means to analyze the nature of cocaine use because of its hidden character and because of the complex cultural elaborations involved. However, if this approach had been the only one used it would not have been possible to obtain information on the extent of cocaine use in the three cities. Furthermore, when researchers exclusively use qualitative techniques, they often have difficulties in both generalizing and validating the results. This is the reason why many field researchers are finding it necessary to link qualitative and quantitative techniques (Mitchell 1966, Korf and Van Poppel 1986, Intraval 1989 and 1991, Grapendaal et al 1991).

4.5   Research design

The research design is based largely on one developed by the Intraval group in drug studies in Rotterdam and Groningen (Intraval 1990). After a period of problem definition and formulation, the research phase of the study began in the spring of 1990. The first three months were devoted to making decisions on the methodological orientation of the project and the techniques to be used. The Intraval design was accepted by the research groups from Barcelona and Turin as the common starting point. This was followed by meetings and discussions between the three parties in which agreement was reached on the following aspects: research methodology, which was to be basically qualitative; the use of snowball and targeted sampling for obtaining the user sample; estimating the extent of cocaine use by means of snowball sampling and the network approach; in-depth interviews; the most important topics of the qualitative and quantitative part of the questionnaire; and the individuals inclusion criterion. At regular meetings, the three research groups discussed the information relating to the specific development of the research project in each city and adjustments were made on the original design.

4.5.1 Data collection

The method used for data collection was a combination of snowball sampling and targeted sampling (Goodman 1961, Biernacki and Waldorf 1981, Watters and Biernacki 1989). As we said earlier, the principle behind snowball sampling is that the initial respondents who are chosen at random are invited to give the names of others who meet the criteria set by the researchers. In turn, these nominees are asked to give the names of others, etc. The inclusion criteria for both respondent and nominee were that they have used cocaine at least 25 times in their life and/or five times in the last six months and are a resident in the city being studied. However, it is virtually impossible to make a random selection of cocaine users as initial respondents since this is a population which is extremely difficult to identify. We therefore decided to combine the technique of a snowball sample with the principle of targeted sampling.

In targeted sampling, researchers deliberately look for respondents whom they have good reason to believe form a reasonable cross section of the research population. Their selection is based on literature and preliminary study. In the course of such a study it may be that certain groups are present only in small numbers or absent from the sample. It is also possible that the researchers discover a group which has never been known. If this happens, an active search is made for respondents in such a group.

4.5.2 Targets: settings of cocaine use

The first phase of the project consisted of an extensive literature study, interviews, formal and informal, with key informants from police services, drug assistance agencies, courts, welfare services, youth work, and the café scene. This preliminary study gave a first general impression of the amount of cocaine use in the cities, delineated observation units and established the first group of informants. An important aim in doing this was to list probable areas of cocaine use. It was assumed that a comparatively large number of cocaine users would be found in the sociocultural settings selected. These settings formed the initial targets of the sample.

In Barcelona, the researchers combined the setting with the type and circuit of most frequent drug use, and included a stratification criterion based on economic status. Five categories were established:
a. Elite: fashion, business, art worlds, etc.
b. New urban middle class: professions, jobs linked to night life, middle-ranks in the fashion business and art world.
c. Young people.
d. Illegal circuits and opiate addicts.
e. Workers: middle status and middle-low status.

In Rotterdam, the targets were based more on sociocultural considerations. Eight settings emerged as probable areas of cocaine use:
a. The hard drug world: heroin users or so-called poly-drug users.
b. Youth circles: young people gathering in places such as youth clubs, community centres, coffee shops selling cannabis, and snackbars. Subgroups were hooligans, and young homeless people.
c. Art, culture and music world: users from the circles of music, the theatre, painting and sculpture, the media etc.
d. The world of fast money: people working in advertising, fashion, and other modern professions.
e. Circles of cannabis users: people who regularly use hashish or marihuana individually and in groups.
f. Illegal and semi-legal circles: juvenile delinquents, drug dealers and prostitutes (male and female).
g. Higher education and university: students and staff at higher education institutes and universities.
h. Sport and fitness world: fitness centres and sports such as basketball, baseball, ice hockey and American football.

In Turin the sampling targets were places where drugs would likely be used. Five different environments of cocaine use were chosen:

a. Places of entertainment.
b. Social and political circles involving young people.
c. The art world and public entertainment.
d. Health Centres for drug addiction treatment.
e. Criminal circles.

These different settings served as the first guideline for assembling data. This definition made it possible to approach potential respondents in a focused way and to establish contacts with various groups of users. As the research progressed, the information obtained was analyzed and used to fix and modify the sampling frame (see also Watters and Biernacki 1989). In the course of this dynamic process, the discovery of new settings, ways of use and new categories of users forced the researchers to open new contact routes. The search for key contact persons and settings was intensified.

4.5.3 Recruitment of respondents

The method used to recruit respondents was a combination of field work and contacts in different institutions, such as (drug) assistance agencies, youth work institutions, and prisons. In Barcelona and Turin the personal networks of the researchers and interviewers were also used. Especially in Rotterdam, much attention was paid to intensive field work, and contacts through advertisements in newspapers were also applied. The field work included becoming acquainted with, and observing, the places where users gather (the so-called field). These activities enabled contact to be established with the cocaine users. Additional information was obtained on the world of the users, with special attention being paid to the following topics:

- the ins and outs of the user world, particularly the manner of obtaining and using cocaine and the degree of use;
- the opportunity structure, which included looking at the places where cocaine is used, and traded, on a small scale.

The field work also provided opportunities to assess the reliability of the information provided by the key informants and respondents.

During the interviews all respondents were asked to give names of other cocaine users. One or two of these nominees were selected (at random) for an interview. This was how the snowball sampling method functioned in practice. The first interviews were the zero-wave of the chain (and the initial sample), the second interviews were the first-wave, and so on.

In Barcelona, the respondents were asked to name all the people belonging to their personal networks who could be interviewed on the basis of the general inclusion criteria. The nominees were classified according to the user categorization described earlier. Two nominees were then chosen at random: one from the category with the most nominees, the other one from the category with the least nominees. In case this first contact failed, the chain was followed with the next selected nominee according to a random procedure. This procedure was repeated until each wave of the chain was completed.

In Rotterdam, the respondent was asked to name ten users known to him in the following five circuits in which cocaine is used: entertainment circuit (cafes, pubs, discotheques and so on), workplace, home circuit, hobby/sports, and hard drug scene. These five circuits were chosen so that respondents would not only name persons in their close environment but also users from more peripheral circles[3]. In this way it was possible to get a reasonable spread of contacts in which cocaine plays a role. Two nominees were selected at random from each circuit using a random number table. The two nominees with the lowest random figures were selected as follow-up respondents. If the respondent said that he was unable or unwilling to contact these people, he was asked which nominees he could contact. If the respondent said that he could not contact any nominee, he was requested to supply information on ways in which new nominees could be contacted.

In Turin, the network of primary contacts was constructed in line with the general inclusion criteria. The respondents were asked to indicate persons among these contacts who could be interviewed. Then they were asked to draw a graph with points corresponding to as many users, or ex-users, as possible, who had not been included in the first listing of primary contacts. Finally, the respondents were requested to indicate those persons who were acquainted with their cocaine use (both among users and non-users) and to draw connection lines between the users. 50% of the nominees were selected from the primary (close) contacts and 50% from the weak contacts. Table 4.1 shows the way the respondents were recruited.

Table 4.1   Method of contact

	Barcelona		Rotterdam		Turin	
	n	%	n	%	n	%
Field work	50	32	31	28	26	26
Advertisements	-	-	19	17	-	-
Drug assistance	9	6	14	13	26	26
Prison	-	-	12	11	1	1
Key informants	27	18	8	7	32	32
Extensions	67	44	26	24	15	15
Total	153	100	110	100	100	100

4.5.4 Interviews

The interview consisted of a questionnaire made up of two parts, one qualitative the other quantitative. The qualitative part was an item list used as the basis for obtaining narrations which reflect the vision of the respondents as fully as possible. After beginning with general topics, the interviewer switched to a more specific level of discussion on the basis of the information provided by the respondent. This process aimed at describing the aspects and events that define the context and meanings of the respondent's relationship with cocaine. It was also a means of testing, by repeated questions, the coherence and credibility of the respondent's information.

The topics of the qualitative section were: the socio-economic background (parental home, childhood, school career, professional career); present lifestyle (daily occupation, expenditure pattern, relationships, leisure activities); first drugs/alcohol experience (how it happened, circumstances, which drugs, age and social circumstances, first cocaine use); cocaine career (pattern, development, method of use, quantities, procurement, problems, abstinence, quitting); setting of use (where, with whom, method of use); function of cocaine (recreational, work, problems, sexual); effect of cocaine (kind of effects, method of use, positive, negative, turnover); consequences of cocaine (in social functioning, work, sexual experience, positive, negative, health, addiction, turnover); income (legal, illegal, reasons for change, how much); criminal offenses (before/after cocaine use, number of times, criminal records, contacts police/ courts); (drug) assistance agencies (contacts, what type, why, career, assessment). The respondent was invited to add items which he considered important. In Barcelona and Rotterdam the interviews were recorded on tape. In Turin, the respondents were unwilling to participate when tape recorders were used and, therefore, the interviewers took extensive notes. Subsequently, the interviews were processed and organized with the help of a word processor programme. Specific characteristics distilled from the in-depth interviews were quantified and statistically analyzed.

The quantitative part consisted of questions with precoded answer categories. These related to the personal network of the respondent, namely cocaine users known to him and contained two groups of questions. The first consisted of seven questions aimed at identifying nominees. Information was requested on: gender; age; nickname; occupation; whether the contact takes place in the context of cocaine; and whether the nominee knows that the respondent takes cocaine. In order to guarantee the anonymity of the nominees without losing the possibility of identifying them, each name was coded. In Barcelona, the code was two first letters of the second family name, the three first letters of the given name, and the first two letters of the first family name. In Rotterdam, the first two letters of first name and surname were used. In Turin the respondents refused to give any initials of the nominees. The second group of questions consisted of seventeen areas of information about the nominee: how long had respondent known him; was respondent present at nominee's first use of cocaine; current cocaine use of nominee; duration of cocaine use of nominee; most important place of use of cocaine of nominee; joint use by respondent and nominee; any link between respondent and nominee in procurement of cocaine. The answers of the quantitative part of the questionnaire were noted during the interview and subsequently processed with the help of a database program.

Additional information was obtained by means of field observations and the writing up of fieldnotes. These data have been processed and arranged with help of a word processor. These fieldwork methods, combined with the questionnaire, provided a wealth of information. Most of these data related to the respondents, but valuable additional data were available about other cocaine users known to the respondents. In Barcelona, information is available on 587 cocaine users: 153 respondents and 434 nominees. The respondents answered all the questions on the quantitative part of the questionnaire concerning nominees. In Rotterdam, there is information on 1,161 persons who use(d) cocaine, comprising 110 respondents and 1,051 nominees. Gender, age and occupation are known of all these persons. With regard to 382 nominees, information is available relating to, among other things, the length of use, place of use, and method of obtaining cocaine. In Turin, there is information about 250 cocaine users, 100 respondents and 150 nominees.

4.5.5 Analysis

A typology of cocaine lifestyles was constructed in order to obtain a description of the nature of cocaine use in Barcelona and Rotterdam. The basis for this was the qualitative part of the questionnaire. In Barcelona, first a characterization of the respondents was made by selecting a range of attributes and variables from the interview. This selection formed the initial profile of the respondent. Subsequently, the characteristics of this profile were analyzed in order to rank them according to importance and select the most relevant. Once the basic type was defined, the respondents were classified. Respondents who did not fit this type were placed in a new category. This process of accumulation and concentration was repeated a number of times. The aim was to obtain maximum homogeneity within the type and maximum heterogeneity between types. Subsequently, the definitive typology was statistically tested for coherency and validity.

In Rotterdam, the analysis technique employed had been developed by the Criminology Institute of the University of Groningen (Janssen and Swierstra 1982, Swierstra and Janssen 1986). Later, this technique was applied, and further developed by, among others, Intraval (1987, 1989 and 1991), Swierstra

(1990) and De Bie and Miedema (1990). First, the individual interviews were analyzed with important elements being distilled from the individual stories. Then the distinctive differences and similarities were looked at. At this stage, the researchers looked at the conspicuous and, at times extreme, respondents who could serve as examples. Based on this, a first rough classification was made of categories of users. First outlines of the dimensions distinguishing the types could be could be obtained. Next, with the help of the so-called minimum-maximum comparison method, the approximate classification was further refined. With this method, groups (types) are formed in such a way that there are minimum differences within the groups and maximum differences between the groups. Respondents placed under one type do not need to score equally on all items but there must be similarities on the most characteristic relevant items (dimensions).

Turin chose an approach different to the one used in Rotterdam and Barcelona. The analysis was done in three steps. First, a definition of patterns of use in each period of use was constructed. Second, a description of cocaine-use styles based on the persistence or variation of pattern of use over time was obtained. Third, the classification of the population of users was made based on the above definitions and descriptions. There was both a qualitative interpretation of the content of the interviews and a statistical analysis of the quantitative variables formed by standardizing answers to the open questions. The statistical analysis was used for supporting the qualitative interpretation and for suggesting explanations.

The quantitative part of the questionnaire was used for the description of the spread and dispersion of cocaine and for estimating the number of users. In order to determine the extent of use, several estimators were developed by Snijders (Frank and Snijders 1992) and one by Del Castillo (Díaz et al 1992). These estimators have been derived from a combination of snowball random samples and the network approach, based on the idea of capture-recapture[4]. Normally, these estimators are based on a randomly selected initial sample but that was not possible in this study. In chapter seven, we describe how these estimators were adapted to the specific selectivity of the initial sample. In addition, the personal networks were analyzed in order to investigate the structure of relationships (contacts) within the group of cocaine users. Likewise, the analysis of the personal networks and the chains provided complementary information which characterized the respondents and described the mechanisms of distribution. The chains were analyzed qualitatively and by means of various statistical techniques such as multi-level analysis, loglinear analysis, principle component analysis, factor analysis, cluster analysis, correlation analysis, Chi-square test, and Friedman and Wilcoxon test.

There were slight differences in the way the research has been carried out. In Barcelona and Rotterdam (in line with the Intraval design) the sampling method based on networks obtained information on the relationships between users: e.g. the degree to which users are aware of each other's use of cocaine, the extent to which the user groups are divided into smaller sub-circuits, etc. In Turin, the approach to the respondents' total personal network ('set') was given greater importance (Barnes 1977). The basic questions were: the size of the primary network (or immediate set), its density, the degree of specialization in the use of cocaine, etc.

The central issue within this apparent discrepancy is a theoretical one. The role attributed to social relationships themselves as frameworks for explanation of the behaviours and lifestyles of cocaine users is critical here. It is an open question to what extent researchers need to know such things as the general individual behaviour patterns and general forms of family and friendship interaction before going on to analyze cocaine use of individuals and their families and friends. These social frameworks in which cocaine use is anchored are comparable to what Epstein (1978) calls 'shared understandings'. A shared understanding is a complex of implicit shared values which render significance to the actions and attitudes of the members of a given social sphere. Respondents in Turin were asked to name the ten most significant persons from the four pre-defined structural environments: friendship, entertainment, work and family. Following this, questions indicating the relationships of these significant persons referred to cocaine were asked. In Rotterdam and Barcelona, the respondents were asked to mention only those other users who fulfilled the general inclusion criteria. In Rotterdam, the people cited were pre-classified into five circuits (entertainment, workplace, home, hobby/sports, the hard drug scene) and the respondent was invited to name up to ten users in each circuit. In Barcelona, the respondents could name as many persons as they wished. These were classified into five pre-established categories based on the different settings of cocaine use and a social stratification criteria (elite, new middle class, working class, young people and students, illegal circuits and opiate addicts).

NOTES

1. Circuit refers to: a group of people in which many relationships exist, or, in network terms, a part of a network with a high density of contacts.
2. A larger initial sample will be needed if the user population consists of a large number of small groups which have no knowledge of each other's existence then if there is a more interwoven pattern in which every user is a friend of a friend of a friend of every other user.
3. There is a danger that otherwise respondents will name only those people with whom they have a close relationship or people they hardly know (for reasons of privacy).

4. The definition of the snowball sample is: The sample is selected at random from the population of cocaine users in the city. This is the initial sample. The people included in the initial sample subsequently nominate people whom they know use cocaine. The nominated persons form the 'first snowball wave'.

PART II

RESULTS

CHAPTER 5

GENERAL CHARACTERISTICS

Francesco Borazzo, Cilia ten Den, Concha Doncel

In this chapter we compare the characteristics of cocaine users in Barcelona, Rotterdam and Turin. The following variables will be considered: background characteristics; effects of cocaine; drug career; characteristics of use; criminality and cocaine trade; problems and contacts with drug assistance agencies. For the comparison, reference has been made to the quantified data derived from the qualitative section of the interviews. A number of quotations from the respondents are also used to illustrate the most important characteristics. In appendix D the corresponding tables are presented.

5.1   Background characteristics

In general, more men than women use cocaine. Two North American studies mention 68% and 56% men compared to 32% and 44% women (Erickson et al 1987, Morningstar and Chitwood 1987). In the study on non-deviant cocaine use in Amsterdam, 60% of the sample consists of men and 40% were women (Cohen 1989). Figure 5.1 shows that also in Barcelona, Rotterdam and Turin there are more male than female cocaine users. The percentage of female users is 33% in Barcelona and 28% in Turin when both respondents and nominees are looked upon. In Rotterdam only 23% are female. This proportion is similar to the female-male ratio among heroin users in the Netherlands (Intraval 1989 and 1991, Korf et al 1990, NVC 1992, Toet and Geurs 1992).

Figure 5.1   Gender (respondents and nominees)

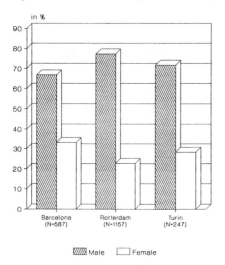

Figure 5.2   Age (respondents and nominees)

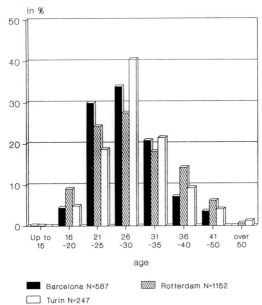

The age of the respondents ranges from 15 to 50 years. In line with the literature, the majority of respondents in the three cities are between 20 and 35

years of age (Spotts and Shontz 1984, Morningstar and Chitwood 1987, Cohen 1989, NIDA 1990). When we add the sample of nominees to the sample of respondents, the frequency distribution within this age-category are as follows: Barcelona 85%, Rotterdam 70%, Turin 81% (figure 5.2).

According to several American studies a large proportion of heavy cocaine users are successful, highly educated yuppies who have no money worries (Spotts and Shontz 1984, NIDA 1986, Morningstar and Chitwood 1987). In both the United States and Europe, however, the view gaining ground is that this no longer corresponds to the real situation (Kozel and Adams 1985, Arif 1987, Cohen 1989). The fact that a number of the cocaine users interviewed in Barcelona, Rotterdam and Turin are also taking heroin (poly-drug users) shows that cocaine use is no longer the preserve of the well-off. In the three cities, all educational levels are homogeneously represented among the respondents. Cocaine users have been found among the employed, unemployed, students and school pupils (see table 5.1). Compared to the other cities a large proportion of unemployed users have been found in Rotterdam (40%). This can be partly explained by the social-welfare situation in the Netherlands according to which unemployment benefits are considerably higher than in Italy and Spain. There is no stigma attached to being unemployed in the Netherlands and the Dutch are more likely to experience periodic unemployment. Furthermore, in the Netherlands there is a special social benefit payment for those with a job but without a regular income. In Rotterdam several respondents, for example artists, receive this kind of benefit: they have been classified unemployed but they do have a job. The number of students or school pupils is approximately the same in each city.

Table 5.1   Daily work

| | Barcelona | | Rotterdam | | Turin | |
	n	%	n	%	n	%
Employed	114	75	43	39	70	70
Unemployed	7	5	44	40	17	17
Student	17	11	8	7	10	10
Other*	14	9	15	14	3	3
Total	152	100	110	100	100	100

* housewives, persons in prison, persons living in closed institutions, etcetera

A classification based on socio-economical status has been used to describe the occupations of the respondents (Domingo and Marcos 1989). In all the three cities, cocaine users are found among professional, intermediate, skilled and unskilled occupation groups (see table 5.2). Only a minority of the respondents are classified in the professional occupation group, the number differing from one city to another. Fewer people with professional jobs were found in Turin than in Barcelona and Rotterdam. In Turin, where industrial production forms the heart of the economic system, a relatively large proportion of users were unskilled workers. This may also be due to the fact that more poly-drug users were interviewed in this city, particularly compared to the number interviewed in Barcelona (41% and 19% respectively). In Barcelona almost half of the respondents had intermediate or professional occupations.

Table 5.2   Occupation

	Barcelona		Rotterdam		Turin	
	n	%	n	%	n	%
Professional	15	13	5	13	2	3
Intermediate	37	32	8	20	14	20
Skilled	55	49	23	59	43	61
Unskilled	7	6	3	8	11	16
Total	114	100	39	100	70	100

With a classification based on socio-economic status much information is lost on the specific characteristics of the occupations. Therefore, the Rotterdam research team developed a second classification, based on type of profession or occupational sector. According to this classification, both prostitutes and sex-club owners have been placed in the category 'sex industry'. Artists, singers, actors, jugglers and musicians have been grouped under 'art/culture'. Figure 5.3 shows the great diversity in the occupations of the respondents. This diversity can be seen in all the three cities. Certain over-represented sectors can however be discerned: art and culture, pubs etc. and the technical sector. Compared to Barcelona, a large proportion of the Rotterdam and Turin respondents were working in the technical sector. This reflects the working class background of these cities. The sector sales is also well represented in Turin. In Barcelona the administrative and media sectors are over-represented, in line with the service sector being Barcelona's main economic feature. Rotterdam is the only city in which respondents were working in the sex industry. This may be due to the special efforts made by the Rotterdam research team to make contacts in this sector and the legal status of sex work in the Netherlands.

Figure 5.3   Type of occupation

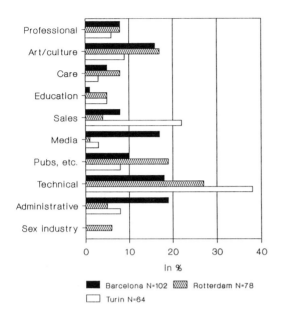

In %

Barcelona N=102   Rotterdam N=78
Turin N=64

5.2   Effects

In all three cities, the respondents describe their first cocaine use as an intense, positive experience, concerning both interpersonal communication as well as their own psychological and physical sensations. In the literature, the sensations are described as a general feeling of well-being and alertness, or in terms of self-confidence and ease of communication. The sensation of the environment taking on intensified qualities is also described. Only a few respondents found the first experience disappointing. They attributed this to the cocaine (bad quality or too little) or to their inexperience.

"It's the usual story of coke that everyone tells, I suppose. It's as if another world is opening up for you, the mist lifts and wow it's crazy! The sun shines brightly and everything is glistening, everything gets an attractive glow and all sorts of ideas shoot into your mind. I really liked that." [R-017][1] "More self-confidence, expansion of your faculties: you can't really make it better, but you can go faster." [T-026] "It makes you feel vital, ready to move and, more than anything, you give up getting into brawls with people about anything. It's perfect at the discotheque,

you're much more sociable, more communicative, more active, more inspired." [B-SF43]

After some time, the majority of respondents still regarded the effects of cocaine positively. However, on occasions, the intake of cocaine can cause negative effects, for example when the quality is bad or the quantity too large.

"There were a number of times that I found the effects less positive. Apparently the coke wasn't good. Once I had a headache and runny nose apparently because the stuff wasn't good." [R-042] "The first time it was pleasant. I have also experienced the negative effects of coke. You get terribly worked up, you don't feel good, sometimes very paranoid and you don't feel safe." [R-062] "Too large quantities make me feel too nervy and my heart beats irregularly and I get very anxious (..) and I see how it makes other people get talkative whereas it often blocks me; I just sit there, it's like I need air." [B-E11] "And one thing that made me really flip, and that made me quite scared, was that when I was 'pumping' with the syringe. I felt as though my heart was going to leap out of my body at three thousand miles an hour. Then it would go to my head and I would fly; I'd see everything so clearly." [B-E16]

Furthermore, some respondents state that the effects of cocaine are diminished when the drug is taken too often.

"If you use it very often then you don't feel it any more. If you take coke every day, you then have to sniff one gram in two goes." [R-057] "If you take it every day you don't notice the effect any more. If you take it a couple of times a week then you still get a kick." [R-062]

5.3 Drug career

The vast majority of cocaine users started taking it before the age of 25 (87% in each city). Figure 5.4 shows that in all three cities the largest group started when they were between 16 and 20 years of age (56% in Barcelona, 44% in Rotterdam and 57% in Turin). In Rotterdam a relative large proportion of respondents started using cocaine before the age of 16 (18%).

There is great variation in the length of use: some respondents used the drug for only three months while others had taken it for more than 15 years. Approximately three-quarters of the respondents have a cocaine taking career of more than five years (77% in Barcelona, 70% in Rotterdam and 72% in Turin). In particular, more than one-fifth of the respondents reported a cocaine career of more than 10 years (28% in Barcelona, 23% in Rotterdam and 22% in Turin). At the time of the interview some respondents were no longer taking cocaine: between one-fifth (Barcelona) and one-third (Rotterdam) of the respondents had ceased taking the drug.

Figure 5.4   Age at initiation

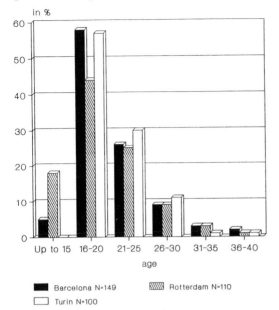

From the literature it is known that the demand for cocaine is more elastic than that for heroin. Individual cocaine consumption is not continuous but intermittent (Peele 1987, Lewis 1989). In the Amsterdam study of Cohen (1989), almost nine out of ten respondents experienced periods of non-use lasting a month or longer, because of lack of money or simply because they did not feel like taking it. In this study, the definition of discontinuous cocaine use is that the respondent has gone for more than one year without a period of use. When this definition is applied, the majority of respondents had a continuous using career. This, however, varies between our cities. In Barcelona only 5% of the respondents showed a discontinuity in use compared to 21% in Rotterdam and 23% in Turin.

> "Inhaling cocaine every weekend is not the greatest aim in my life. Therefore there are periods when either I don't feel like it or else I haven't got any money and ... well, I'm not that keen on it." [B-SF48]
> "Intervals even of one year. Not because I had problems, but rather because I couldn't keep it up with the money I had." [T-029]

The literature states that in general cocaine users also take other drugs, usually alcohol and cannabis. Virtually every cocaine user in the United States has had some experience with marijuana or hashish. Sedatives and sleeping pills are mainly used as calming agents after a cocaine run. Only a minority of cocaine users in the United States have experience with heroin and/or other opiates (Spotts and Shontz 1980, Siegel 1984, Kozel and Adams 1985, Erickson et al 1987, Erickson and Murray 1989). Table 5.3 shows that, in accordance with the literature, the majority of cocaine users in Barcelona, Rotterdam and Turin have had experience of various other drugs. There are hardly any respondents who used only cocaine (less than 5%). The other kinds of drugs (often) used are cannabis, amphetamines, MDMA (XTC), mushrooms and LSD. Sedatives and sleeping pills are rarely mentioned. Some respondents in the three cities also used, or had used, heroin; 19% in Barcelona, 33% in Rotterdam and 41% in Turin. In these cases, cocaine and heroin are often taken together (i.e. the so-called speedball) or heroin is used as a calming agent after a cocaine run. In Rotterdam and in Turin, 25% and 30% of the respondents respectively were already using heroin before starting to take cocaine[2]. In each city respondents were found who started using heroin after already using cocaine. Cocaine functioned as a gateway drug to heroin for these users.

"I use also hashish and XTC ... but it's coke that gives me force. It's a super-vitamin having immediate effect." [T-051] "Always drunk alcohol. We blowed (smoked cannabis) too. I used LSD once in my youth, I've also taken speed and marihuana, also once mushrooms." [R-005] "I used to be addicted to valium and librium. And at a certain moment I didn't take them any more because of coke. And I really liked that, of course." [R-101] "Whenever I was going really heavy on cocaine and needed to relax, then I'd use heroin." [B-DC3]

Table 5.3   Kind of drugs throughout life

	Barcelona		Rotterdam		Turin	
	n	%	n	%	n	%
Only cocaine	-	-	2	2	4	4
Also others (but not heroin)	123	81	72	65	54	55
Also heroin	28	19	36	33	40	41
Total	151	100	110	100	98	100

The substance that is named most frequently in relation to cocaine is alcohol. Cocaine dependence often goes hand in hand with alcohol dependence (Anthony 1991). The respondents from Barcelona, Rotterdam and Turin also state that alcohol, both before and during their cocaine use, is the substance that is taken most often. Moreover, many respondents stated that the intake of alcohol tends to increase when cocaine is consumed.

"I can drink ever such a lot without getting a headache, whereas as if I drank the same amount without taking any cocaine I'd be falling around

all over the place." [B-SF47] "Alcohol and coke match really well: alcohol is exuberance, coke is clearness and lightness." [T-079] "We started drinking more whisky instead of beer. It tasted much better with it. With coke, I could also drink more. You don't get drunk at all." [R-103]

5.4   Characteristics of use

The pattern of use over time has been drawn for each respondent on the basis of in-depth interviews with 363 respondents in the three cities. Both frequency and consumption level were taken into account. An examination distinguished six basic patterns of use in the three cities (see also Cohen 1989). The patterns are presented graphically in appendix E.
1.   Increasing: the consumption level gradually rises.
2.   Decreasing: the consumption level gradually falls.
3.   Same level: the level remains the same.
4.   Peak: a gradual rise in intake is followed by a gradual decrease.
5.   Twin peaks: the consumption level varies and is characterized by different peak intake levels.
6.   Discontinuous: periods of drugs use are interspersed with periods of abstinence[3].

From figure 5.5 it becomes clear that the most common patterns of use are those in which one or more peaks can be recognized (peak 35% and twin peaks 14%), or in which no evolution can be seen (same level 18%).

Figure 5.5   Patterns of use

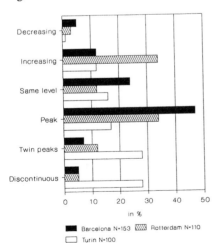

In a large group of respondents from Rotterdam (34%), however, a continuously increasing pattern, typical for heroin users, was found. In contrast, 28% of the Turin respondents showed a discontinuous pattern, whereas 25% of the Barcelona respondents took cocaine at the same level throughout their using career. In general, these results are comparable to the results of several sociocultural studies carried out in the United States and the Netherlands; Chitwood (1985), Murphy and co-workers (1989) and Cohen (1989) also found a high prevalence of patterns characterized by variation in use level and of patterns characterized by the same level of use over time.

In analyzing the data on the characteristics of use, a distinction has been made between the initial period of use, the present or most recent period of use, and the period of heaviest use. The initial period of use is defined as the first period in which the respondent occasionally took the drug. The heaviest drug taking period is that in which the respondent took cocaine most frequently and/or consumed the greatest quantities. For a number of respondents the period of heaviest use coincides with the initial period or with the period of current/last use. For another group of respondents their heaviest drug taking phase constitutes an interim stage. When examining the different patterns of use, we discover that for the majority of respondents in the three cities the period of heaviest use is the interim period.

"The frequency has decreased much, the occasions are more and more sporadic. It is more and more a choice of mine." [T-005] "It varies, a period of not taking it and then I start again. An average of once a month, one small line." [R-097] "Between 25 and 30 (years of age) the consumption rate is on the increase. In the beginning (...) just a little line on Fridays (...) and then after a month or so, every day. So, at say 25 or 26, for a year I was consuming daily and then only in seasons; only in the summer and at Christmas, say." [B-CT11]

To describe the social surroundings or the location in which cocaine is taken, the following circuits have been defined: entertainment circuit (cafes, pubs, discotheques and so on), work circuit, intimate circuit, hobby/sports circuit and the hard drug scene[4]. The most recurrent circuits of use during the initial period are the intimate circuit (Rotterdam and Turin) and the entertainment circuit (Barcelona). We found certain differences between cities as regards the period of heaviest use. In Barcelona there is a diffusion into several other circuits: entertainment, intimate and work. In Rotterdam, the hard drug scene becomes more important. In Turin, the entertainment circuit becomes more important (see table 5.4). In general, we can say that the entertainment circuit is most important in Barcelona and the intimate circuit in Turin. In Rotterdam, the entertainment, intimate and hard drug circuits are equally important. In no city were the work and the hobby/sports circuits important locations in which cocaine is consumed.

"Like for most people, initiation takes place in a group. I don't know anybody who consumed it the first time out on their own. It was at a party (...) with people I feel at home with: they are my friends." [B-SF44]
"I take cocaine on a relaxed evening with a few of friends. At most six of us. Good food, with wine. And, just as I said, between the dessert and the coffee with calvados the mirror is presented with a line of coke." [R-069]
"When we go out, we use cocaine in the car or in a doorway. At the beginning it was more at someone's home and, later on, more when we were out." [R-064]

Table 5.4   Most important circuit in first (F) and heaviest (H) period of use (in %)

| | Barcelona | | Rotterdam | | Turin | |
	F	H	F	H	F	H
Entertainment	46	62	27	29	11	36
Work	10	8	5	6	3	6
Intimate	37	2	47	33	83	52
Hobby/sports	1	-	2	1	-	-
Hard drugs	6	8	18	30	3	6
Three equally important circuits	-	20	-	1	-	-
Total %	100	100	100	100	100	100
Total N	144	141	110	106	100	99

Cocaine is administered in many different ways. It can be sniffed, injected intravenously, or used by way of free basing. It is rarely smoked in a cigarette, if only because this is an expensive and wasteful way of using the drug. The great majority of cocaine users, most certainly outside the world of the heroin addict, sniff the drug (Cohen 1989, Van Hunnik 1989). The data from Barcelona, Rotterdam and Turin show that sniffing is indeed the most common way cocaine is used in these cities, although some of the respondents use cocaine in more than one way. In most of these cases sniffing is combined with another way of use, usually smoking or basing. For the majority of the respondents, however, sniffing is the most important way cocaine is used in every period of use (first, last and heaviest). Corresponding to the literature, there is a clear difference in the way cocaine is administered by respondents who also used, or had used heroin, (heroin users) and respondents who never used heroin (non-heroin users) (see table 5.5).

Table 5.5   Most important way of use during heaviest period, for heroin users and non-heroin users (in %)

	Non-heroin users			Heroin users		
	BCN	RT	TRN	BCN	RT	TRN
Sniffing	97	86	98	32	17	74
Injecting	-	-	-	64	49	26
Basing	3	10	-	4	17	-
Others*	-	4	2	-	17	-
Total %	100	100	100	100	100	100
Total N	125	74	60	28	36	39

* smoking in cigarette and chasing the dragon
BCN = Barcelona; RT = Rotterdam; TRN = Turin

Almost all non-heroin users sniff cocaine, while heroin users also inject cocaine especially in the period of heaviest use (64% in Barcelona, 49% in Rotterdam and 26% in Turin). Free basing is a way of use particularly common in Rotterdam among both heroin users and non-heroin users. In Rotterdam there is the greatest variation in the way cocaine is taken: sniffing, injecting, basing, and chasing the dragon[5]. Cocaine self-administration by chasing the dragon is unique for Rotterdam. In Barcelona cocaine is self-administered by way of sniffing, injecting and basing and in Turin only by sniffing and injecting. It is remarkable that only a few heroin users in Turin inject cocaine compared to those in Rotterdam and, in particular, Barcelona. Summarizing, we can say that there are clear differences in the methods of administering cocaine in the three cities.

"What I like about sniffing coke is the ritual you go through with a group of people. Just like passing the joint in the old days. I like that; it makes it a cosy happening." [R-100] "I don't only snort it, sometimes I wet a cigarette or put it onto my teeth." [T-043] "Snorting, basing and inject-ing. You can make a stairway, in fact. Snorting is the most relaxing, on foil (chasing the dragon) you get worked up more quickly, basing is again more intensive and with a shot you have a real explosion. But the feeling is really only for a minute and then you feel yourself getting cold blooded. With basing I find it very difficult to cope with the after-effect. Then you feel very unsure of yourself. That is that really paranoia feeling. Some people just allow it to fade but I can't wait so long. I make sure that I have some smack." [R-085] "Chasing the dragon, you light up while its lying on the foil. You have a pipe in your mouth and you hold the lighter under it. Basing isn't worth anything. With basing you have a glass in front of you with a bit of ash at the bottom. You fold some foil

around this and put an elastic band around it. You make a couple of
holes on one side. By then pressing your lighter on the ash and your lips
over the holes. The smoke which comes into the glass you suck out of the
glass again. But that works only five minutes. I have never shot." [R-023]

The frequency of cocaine use can vary from incidental to daily use, indicating the role the drug fulfils in the life of the user. In the American literature (Siegel 1984, Stone et al 1984, Inciardi 1986), a distinction is made between:
- experimental use: no pattern can be discovered (yet);
- social-recreational use: approximately ten times a year cocaine is used at a party or another social event;
- routine use: usually at work, small amounts are taken on a regular basis;
- intensive use: use takes place on a daily basis, at work or out of boredom;
- compulsive use: the use of cocaine takes place under all circumstances, at all times.

Table 5.6   Frequency of use during heaviest period, for heroin users and non-heroin users (in %)

| | Non-heroin users | | | Heroin users | | |
	BCN	RT	TRN	BCN	RT	TRN
Daily	43	38	27	82	86	49
Weekly	32	38	34	7	8	36
Monthly	10	17	34	7	3	15
< than monthly	15	7	5	4	3	-
Total %	100	100	100	100	100	100
Total N	124	74	59	28	36	39

The frequency of cocaine use also varies greatly among the respondents of Barcelona, Rotterdam and Turin, from a few times a year to daily use. In the period of heaviest use half of the respondents in Barcelona (50%) and Rotterdam (54%) and 39% of those in Turin used cocaine daily (at least four times a week). Compared to the initial period of use, monthly use (once or twice a month) and less than monthly use decreases sharply in this period. Non-heroin users in all the three cities tend to take cocaine on a monthly or less than monthly basis. Table 5.6 shows that most of them use it weekly (once or twice a week) or daily in the period of their heaviest use. Heroin users tend to use cocaine on a weekly or daily basis: in the period of heaviest use the majority takes cocaine daily. Compared to the respondents in Barcelona and Rotterdam, the Turin respondents take cocaine less frequently.

Next to frequency of use, there is also great variation in the level of use (amount of cocaine per month). In all the three cities, there are respondents who take less than half a gram of cocaine per month and those who take more than 50 grams per month. The level of cocaine use of non-heroin users in all the three cities is lower than that of heroin users (see table 5.7). This is true for all the different use periods. The Turin respondents take cocaine less frequently and have a lower level of consumption than the respondents of Barcelona and Rotterdam.

"Every two months a gram. That costs 200 guilders ($100). I used it when it was convenient and when I had the money. I took a line at home and then went out on the town. There was a period that I didn't take anything since I didn't have the money." [R-060] The minimum, hardly anything at all, maybe a little line or two a night, at weekends. The maximum was four grams a day. For me the actual effect of the drug started to lose it's importance, the important thing was the fact of using it; the important thing was to drug yourself." [B-SC23] "When I was around 20 I was injecting it all the time and there were times that I was even injecting about two grams a day. One day I jabbed myself four grams... that was... well, for jabbing yourself four grams, you have to make a lot of holes." [B-E16]

Table 5.7   Level of use (amount per month) during heaviest period, for heroin users and non-heroin users (in %)

	Non-heroin users			Heroin users		
	BCN	RT	TRN	BCN	RT	TRN
< 0.5	12	15	6	-	4	-
0.5 - 1	12	7	9	-	4	-
1 - 2.5	10	5	23	4	-	16
2.5 - 5	16	7	11	11	-	5
5 - 10	18	20	13	14	7	19
10 - 15	19	24	2	32	63	-
> 15	13	22	36	39	22	59
Total %	100	100	100	100	100	100
Total N	124	74	53	28	36	37

5.5   Income, crime and cocaine trade

As far as income of the respondents is concerned, a distinction has been made between income acquired legally, semi-legally or illegally. The category legal has been reserved for respondents who acquired their income exclusively in a legal way (wages, salaries or social benefits). When respondents (also) acquired an income from a job on the side or prostitution they have been categorized as semi-legal, when they (also) acquired income by criminal activities (such as offenses against property or drugs dealing) they have been put into the category illegal. In the first period of use, 53% and 66% of the respondents from Rotterdam and Turin acquired their income exclusively in a legal way; 28% and 17% acquired their income in an illegal way[6]. In both cities the percentage of those with illegal income rises in the period of heaviest use, in Rotterdam to 43% and in Turin to 24%[7]. During this period, almost the same number of respondents in both cities acquired their income in a semi-legal way (18% in Rotterdam and 14% in Turin).

Consciousness-influencing substances such as cocaine which attract publicity are readily blamed for criminal behaviour. However, it is difficult to prove scientifically that such a connection can be made for specific substances (Watters et al 1985, Leuw 1988). Furthermore, people who use only cocaine are not easily traced by the police, so that accurate data are not available. At the level of the user, a distinction can be made between criminal activity surrounding small-scale trading, income generating crime, and other crime-reinforcing effects of cocaine. Small-scale dealing takes place partly in the same circles as the acknowledged problem drug heroin. Criminal behaviour is normal in that world. In addition, there exists a separate dealer circuit for the socially integrated cocaine user. It is possible that there are fewer links in the chain between the importer and consumer of cocaine. This could mean, paradoxically, that individual dealers, including those in socially integrated and elite circles, have far more serious criminal connections (since they are dealing in larger quantities) than the small-scale heroin dealer (Lewis 1989). As far as acquisition crime is concerned, cocaine costs at least as much as heroin on the black market. Furthermore, since cocaine is an expensive habit, in the periods that a user is taking an above average amount of cocaine he may need to generate more income (Van Limbeek 1986, Grapendaal 1989). Other criminal consequences are in the main linked to the uninhibiting effect of cocaine: macho and dare-devil behaviour. Analysis of incidental cases has shown that cocaine in combination with alcohol can play a major role in violent crime (Van Epen 1988, Budd 1989).

Regular use of cocaine is, however, not necessarily accompanied by criminal behaviour. One third of Cohen's respondents have at some time been connected with criminal activities. Excluding the offence dealership, the percentage drops to a low of 15%. Only a negligible percentage has been involved in ten or more criminal acts other than dealing (Cohen 1989). Also in Barcelona, Rotterdam and Turin only a minority of the respondents appeared to be criminally active in the initial and last period of cocaine use, ranging from 10% in Barcelona to 35% in Rotterdam. The larger part of the criminality consists of property offenses and drug trafficking. The number of respondents who are criminally active increases in the period of heaviest use to 19% in Barcelona, 48% in Rotterdam and 33% in Turin. The differences between the cities might be explained by the fact that more heroin users have been interviewed in Rotterdam and Turin. Figure 5.6 shows that cocaine users who also use(d) heroin are by far more criminally active in their period of heaviest use than cocaine users who have never used heroin. This applies for every use period.

Figure 5.6   Criminal activities in the period of heaviest use, for heroin users and non-heroin users

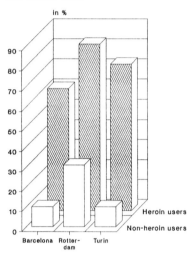

Figure 5.6 suggests also that the Rotterdam respondents are more liable to commit a criminal act than the respondents in Barcelona and Turin. This could be explained by the fact that a relatively large proportion of the Rotterdam respondents, both heroin and non-heroin users, have been interviewed in jail. In addition, it was noticed that Rotterdam respondents were more willing to openly discuss their criminal activities than the respondents in Barcelona and Turin.

"I would never do anything bad to get it. But I do spend a lot of money on coke." [R-050] "Coke covered the trouble of heroin: very useful if you want to peddle. It helps you not to be caught by the police." [T-055] "I've also pushed the drug. At the beginning I was afraid of being caught on the frontier, then I found someone who, in exchange for a few doses, let me pass." [T-048] "Right at the beginning I had money problems. At a certain moment I couldn't pay the rent from my social security benefit. I then soon started breaking in, forging cheques and the like. When later I

really got into the underworld I didn't really have money problems any more. I then did everything which God has forbidden. Smuggling, courier work, etc. All that had no direct link to cocaine, but on the other hand it did. It was simply intertwined with coke. It was then already that I was addicted to coke. It was available all the time, but I was expected to do something for it (criminal activities)." [R-101] "Whenever people get into these dynamics of consuming the drug every day -what a wow!- they have certain expenses that lead them inevitably into having to start selling the drug themselves, it's the only way out... with heroin there are other alternatives, like small thefts. But with cocaine, small thefts are not enough, and you have to make big ones or dedicate yourself to selling. The two or three people I've known who were injecting themselves with cocaine are people like that." [B-P16]

As far as cocaine trafficking is concerned, it should be noted that 27% Barcelona, 31% Rotterdam and 36% Turin respondents have been involved in cocaine trafficking in one way or the other. Not all respondents involved in the market can been considered dealers. Particularly in Barcelona and Turin, a substantial proportion of those involved merely resell small amounts to good friends (Barcelona 50%, Rotterdam 26% and Turin 40%). This is also reflected in the circuit of procurement: the people from whom cocaine is most likely to be obtained are good friends (especially in Barcelona and Turin), followed by house dealers who do not sell heroin (see table 5.8).

Table 5.8   Most important circuit of obtainment of cocaine

| | Barcelona | | Rotterdam | | Turin | |
	n	%	n	%	n	%
Entertainment	4	3	8	8	1	1
Intimate	59	42	27	27	43	43
Home dealer, no heroine	50	36	29	30	19	19
Home dealer, + heroine	4	3	24	24	15	15
Market	21	16	11	11	22	22
Total	138	100	99	100	100	100

A significant number of the respondents in Barcelona and Turin obtain cocaine directly from the market, whereas the heroin dealer who is also selling cocaine is more important to the Rotterdam respondent. Considering the importance of the entertainment circuit as the location in which cocaine is consumed, it is remarkable that only a minority of the respondents buy their cocaine mainly in this circuit (from 1% in Turin to 8% in Rotterdam).

"I've got a secret diary including the names of reliable people who can buy for me. I prefer to pay more, but not to run any risk getting involved with infamous people. I trust only a few persons, who would have nothing to gain by talking.." [T-054] "The first time I got it from friends and later on I often bought a pack for a hundred guilders ($ 50) together with a mate. The dealer I usually got it from dealt in cocaine, hash and weed." [R-080] "I could buy both there (heroin and cocaine). A dealer who sells only coke or smack isn't worth anything. He must always have both." [R-085] "I very often have clients for a dealer. I often fetch it for friends." [R-057]

5.6 Problems and contact with drug assistance agencies

Approximately half the respondents in Rotterdam (55%) and Turin (47%) and three-quarters in Barcelona (75%) admit having (had) problems related to the use of cocaine. This concerns a great variety of problems: from a hangover the next day (psychological and/or physical), lack of money, to aggressive feelings and craving. In Barcelona and Turin, physical problems are most often mentioned while in Rotterdam psychological problems (for example feelings of guilt and even symptoms of paranoia) predominate. The small subgroup of respondents in the Amsterdam study who volunteered for examination also revealed no serious physical problems, whereas some defects in their mental functioning were observed (Cohen 1989). In Barcelona 15%, Rotterdam 25% and Turin 7% of the respondents have severe problems due to cocaine use: they mention physical, psychological, social as well as economic problems or describe themselves as addicted to cocaine. Figure 5.7 shows that in all the three cities severe problems due to cocaine use are more frequently mentioned by heroin users than by non-heroin users. The Turin respondents appear to have less severe problems due to cocaine than the respondents from Barcelona and Rotterdam. In section 5.4 it has already been noted that the Turin respondents use cocaine less frequently and at a lower level than the respondents in Barcelona and Rotterdam. Furthermore, injecting and basing are not popular in Turin. In the following chapter the relation between way of use, frequency and level of use and the occurrence of problems will be discussed more deeply.

"You get up in the morning and there are no after effects, nothing at all. (...) The worst thing about it is that it takes you ages to go to sleep if you've had a lot, because then you'll be too excited, but that's all." [B-SC8] "In the morning I get a sort of financial gripe, purely from the fact that I wasted too much money. That it just slipped through my fingers while there's all sorts of other things I could have done with that money." [R-030] "A negative effect is my lungs. Not only with basing. At a certain moment when you sniff a lot you get trouble with your lungs.

Yes, you run the risk of catching a cold and that you get a chronic cold..." [R-017] "After a time like that (a period of more intense use) you have a period that you are down and depressive. Particularly after a longer period like that it can be quite severe." [R-066] "The great chance occurred when I shot it. From then on, I began using a lot (...) Then I couldn't stay even half an hour without using, I was caught by such anxiety.." [T-072] "Coke made me arrive where I did not go by myself. Yes, in the end I thought only of the drug: I broke with everybody, my family; my partner was already in prison." [T-074] "If I really went completely overboard, that's when I used a lot, then I got a persecution complex. Then I looked thirty times over my shoulder in twenty metres when I walked home from the place I worked. That was horrible."
[R-092]

Figure 5.7   Heroin users and non-heroin users with at least three types of problems or say they are addicted to cocaine

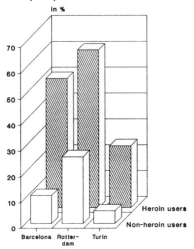

Despite the occurrence of problems, contacts with drug assistance agencies are not common in Barcelona and Turin. In these cities only 8% and 16% of respondents, respectively, made use of the services offered by these agencies. In Rotterdam, a larger number of respondents (38%) contacted the services. The majority of these respondents are poly-drug users for whom a wide variety of services are available in Rotterdam.

NOTES

1. [R-]: quotation respondent Rotterdam; [T-]: quotation respondent Turin; [B-]: quotation respondent Barcelona.
2. No data on Barcelona available.
3. As far as the discontinuous pattern is concerned, it should be noted that many of the respondents passed through one or more phases of total abstinence from cocaine (see Kaplan et al 1992). In Rotterdam and Turin, it is only when the drug taking pattern is characterized by repeated periods of use and non-use of cocaine that it has been qualified as discontinuous. In Barcelona, a discontinuous pattern is characterized by one or more periods of non-use of over a year.
4. There is a difference in the way the locations of use have been classified in the three cities. Barcelona and Rotterdam focused on the social surrounding, whereas Turin focused more on the specific location in which the use of cocaine took place. When poly-drug users take cocaine in an intimate surrounding with friends who also use heroin, this has for instance been classified as use in an intimate circuit in Turin, whereas in Barcelona and Rotterdam this has been classified as use within the hard drug circuit.
5. Some of the Rotterdam respondents who also use(d) heroin take cocaine by the way of use of 'chasing the dragon' (17% in the period of heaviest use). Cocaine is heated on silver paper and the fumes are inhaled with the aid of a (cardboard) tube. In the Netherlands this is the most common way of taking heroin (Intraval 1991).
6. No data on Barcelona available.
7. It was noted that the Rotterdam respondents were more willing to openly discuss their illegal activities than the respondents in Turin. This may, in part, account for the differences in income found between the Rotterdam and Turin respondents.

CHAPTER 6

TYPOLOGY

Edgar de Bie, Mila Barruti, Maria Grazia Terzi

An important feature of the research question in this book is to gain insight into the nature of cocaine use in three different European cities. To this end, extensive interviews were held with more than 360 cocaine users. In Barcelona 153 cocaine users were interviewed, in Rotterdam 110, and in Turin 100. The in-depth interviews were the basis for the qualitative analysis that forms the central feature of this chapter. Before looking at the results, we will focus briefly in the next section on the different methods of typology construction that have been applied in the three cities. After that, on the basis of a comparison of the types of cocaine users identified in each city, a more general typology is constructed.

6.1 Methods of typology construction

In general, Barcelona and Rotterdam used a similar approach, differing only in detail. The use of cocaine was linked to the broader concept of lifestyles. Turin has chosen a different approach. Use of cocaine was linked to the relationship between user and substance in three different periods in the cocaine career. These differences are partly due to the composition of the local teams. In Turin the public health background of the research team is reflected in the typology construction. The Turin typology focused mostly on those aspects of cocaine use that provided insight into feasible preventive and therapeutic interventions. The teams of Barcelona and Rotterdam consisted for the most part of social scientists who tended to look at the global nature of cocaine use, including both pathological and social aspects, and from that basis to necessary and possible aspects of prevention and intervention.

In Barcelona, the basic material was formed by the 153 in-depth interviews (Díaz et al 1992). To construct the lifestyle typology, first a characterization of the respondents was made by selecting a large number of attributes and variables from the interview. This selection gave an initial profile of the respondents. Subsequently the characteristics of this initial profile were analyzed in order to establish their degree of relevance. Then a selection of the most useful of them was done. On the basis of this filtering process, the following characteristics were chosen as dimensions to classify the respondents: aim of cocaine use, context of cocaine use, lifestyle, and importance and meaning of cocaine as stated by the respondents. In the analysis, the respondents 'itinerary' (diachronic analysis of their life histories focused on cocaine use) has also been taken into account. After this, a basic, clearly delineated, type was defined. The respondents who did not fit this type were placed in a special category which in turn was examined for group similarities. If enough similarities were established, a new type was defined. Once again the respondents who did not fit in this type were placed in a separate category. This process of accumulation and concentration was successively repeated. The provisionally defined types were tested, and, at the same time, improvement of the inclusion criteria and the type margins (maximum homogeneity within the type and maximum heterogeneity among types) was obtained. Once this provisional typology was constructed, it was tested and adjusted taking into account other characteristics of the respondents which were not chosen as dimensions for the initial construction. During this process of analysis, refining and defining the typology, the interviews were constantly referred to and the respondents and types re-evaluated. Subsequently, the definitive typology was subjected to statistical tests to contrast and validate it descriptively.

In Rotterdam, the basic material for the qualitative analysis was formed by 110 in-depth interviews (Intraval 1992). The first phase entailed a thorough reading of the interviews. Two or three striking characteristics were noted for each respondent; for example, a specific way of use, an occupation, or some other noticeable feature. This provided an initial means of getting to know the respondents, to be able to recognize them and tell them apart. The most relevant items became apparent during this stage. The second stage involved making profile sketches of the individual respondents. Attention was paid to biographical characteristics such as user's background, occupation, lifestyle, drug use prior to using cocaine, the initiation into cocaine, further development and duration of use, the most frequent setting for use, contacts (if any) with the law or agencies offering care and treatment. In addition, attention was paid to more implicit information inherent in various remarks such as reasons for use, reasons for changing the pattern of use, the role cocaine plays in the respondent's life, the meaning attached to use, opinions concerning the addictive nature of cocaine and a respondent's dependence on the drug. This stage incorporated a first interpretation of the empirical material. As such, the profile sketches were more than just summaries of the in-depth interviews. In the third phase, categories were formed by sorting striking and extreme respondents together with other respondents who resembled them. In this way, a tentative classification was created. Similarities and differences in the aspects of these categories were noted. The procedure no longer referred to individual differences and similarities, but to similarities within and differences between categories; i.e. the procedure moved from the individual to categorical level of analysis. Following this, the categories were once again compared with each other. It was noted which were similar and which not, and whether a common denominator could be discerned. This did not refer to simply recording and counting aspects, but rather to the connecting of the various aspects. By merging some categories and splitting others, new categories were formed which were more clearly distinguished from one another. The differences between categories were maximized. In this way, distinguishing dimensions became more clearly visible. Internally, on the other hand, sufficient similarities had to remain to justify the use of the term 'category'. This was accomplished by a process of trial and error. An example of a poorly-chosen dimension was 'instrumentality of use'. According to this dimension, poly-drug users who inject daily and use cocaine solely to pep themselves up fell into the same category as people who use the drug once a month to allow them to keep going during an intensive weekend. Cocaine has an instrumental character for both categories. Therefore, such a dimension in itself formed too rough a distinction and was rejected because there was no longer a maximum similarity within a single category. This third step of merging and splitting was repeated until the differences between categories were as broad as possible and the similarities between categories as close as possible (the maximum-minimum comparative method)[1].

Turin chose a different approach to Rotterdam and Barcelona (Merlo et al 1992). Turin's applied method was twofold: on the one hand a qualitative interpretation of the content of the interviews and on the other a statistical analysis of the quantitative variables in the structured part of the interviews. The statistical analysis was then used for supporting the qualitative results, as well as for suggesting new possible explanations. This analysis finally leading to a classification of users, took place in three distinct steps. The first step was a distinction and definition of 'styles of use'[2]. The data to describe these styles were obtained and analyzed on the basis of statistically significant correlations found between specific descriptive variables. These variables are: circuit of use, route of ingestion, presence of criminal activities connected with cocaine, sources of income, use or non-use of heroin before or during cocaine use. In this way different styles could be distinguished, which could be characterized as follows:

- Intimate. The persons who use cocaine according to this style are persons who do not commit any crimes connected with cocaine use, have legal sources of income, do not inject cocaine, and use it in friendship or entertainment circuits.

- Work. This is the style of use of those who do not commit any crimes connected with cocaine use, have legal sources of income, do not inject cocaine, and use it not only in friendship or entertainment circuits but also in the work circuit or tied to work in some way.
- Hard. Those who use to commit crimes that they define as being connected with cocaine, obtain their income through proceeds of illegal activities, use cocaine intravenously, and have used heroin either before or during cocaine use.
- Others. The heterogeneous group of the modalities of use that did not fall within the intimate, work and hard styles.

In the second step, this synchronic approach was placed in a diachronic perspective. By looking at these styles of use over the course of time, it was seen that each cocaine user can either fluctuate among different styles from one period to the other or maintain the same style over time. The distinguished styles of use were related to three different periods of use (the first, the heaviest, and the last or current period). Thus, a description of the persistence and variation of these styles over time, the so-called cocaine use styles was produced. The last step entailed the classification of the sample of cocaine users by the above dimensions; styles of use and persistence/variation over time.

6.2   Types and classes of cocaine users

In this section attention will be paid to the different types and classes of cocaine users which have been discovered in the three cities. In diagram 6.1, an overview is presented of all the types and classes.

Diagram 6.1   Types and classes of cocaine users

Barcelona	Rotterdam	Turin
Social	Burgundian	Intimate Group
Circumstantial	Experience	Work Group
Situational	Situational	Hard Group
Elitist	Distinctive	Changing Group
Commercial	Hedonist	
Dysfunctional	Routine	
- Pure cocaine addict	Poly-drug	
- Cocaine addict ex-heroin addict	Cocainists	
Heroin addict		
Former heroin addict		

For reasons of space, only brief descriptions of the different types are given here. Only the most central features have been highlighted. In the local reports of Barcelona and Rotterdam, there are extensive descriptions of their types (Díaz et al 1992, Intraval 1992).

6.2.1 Barcelona

During the construction of the typology of Barcelona, different techniques were used. Each respondent was classified in accordance with the different features comprising his profile. These included the following: the purpose and function of their use of cocaine, the context of use, the lifestyle, and the importance that was given to the drug by the respondent. Also the respondent's 'course' was taken into account (a diachronic analysis of the biography focused on the use of cocaine). This typology was countered and validated by using statistical techniques such as cluster analysis, multiple correspondence and the classification analysis in accordance with criteria[3]. Finally, eight types of cocaine users could be distinguished.

Social type

"Never out of the context of the weekend. I never think about using cocaine in the middle of the week, if there are no festivities or parties. Cocaine only has sense at night, in a party, when you go out, not on a normal working day. I can not imagine myself sniffing in the toilet of my house." [B-SF49]

For this type, the use of cocaine takes place in a party ambience and is always directly related to social gatherings. Use outside this context is evaluated negatively. What is important is the party itself and the social interaction that take place there. If there is cocaine the ambience gets better, if there is no cocaine it does not matter, it is not indispensable. The role of cocaine is, for instance, not as important as that of alcohol and other drugs (especially, hashish). Heroin, on the other hand, is expressly rejected. People do not dare to experience this drug. The pattern of consumption stays stable over time. Cocaine is used sporadically and the amount used is, generally, less than 0.5 gram per month. The route of ingestion is always intranasal. The vast majority have no problems related to cocaine use and none of them has had contacts with the social-health services. Nor has any respondent of this type committed any criminal activity in relation to cocaine. This type consists of 52 respondents of whom 24 are women. The average age is 27 years.

Circumstantial type

"What I consume varies according to the economic situation I'm going through, but at weekends it ranges between one gram and a quarter. Whatever you can, whatever you can get at the weekend (...) During the holidays it's more or less every day, well every day, to be honest." [B-CO23]

The context of use of this type is still the party. The role of cocaine is, however, more important and significant. There is a clear association between a party and cocaine (sometimes, if there is no cocaine, there is no fun). Frequently the use of cocaine goes together with a high consumption of alcohol. Within this type, conscious use intended to counter the party's negative effects (tiredness, hangover), can be found. For this reason cocaine may also be used at home or in a working context. In such circumstances, the use of cocaine becomes more concealed. In situations of personal and emotional crises, during periods of intense work pressure or occasions that require greater performance, the level of use can temporarily increase. This type consists primarily of workers and middle managers. None were addicted to heroin and they always take cocaine intranasally. Although there are problems derived from the use of cocaine, no one has had contact with the social-health assistance network. None have carried out criminal activities related to the use of cocaine. This type consists of 23 respondents. Twelve respondents are men and the average age is 28 years.

Situational type

"I have never been really hooked, (using cocaine) every day. Every day ... maybe a year ago, possibly, when I was working at night as a waitress." [B-SC18]

The characteristic feature of this type is that using cocaine is mainly related to the work context; a context in which cocaine is available and where the use seems to be linked to the performance requirements of the context itself. This relates specifically to certain circles such as night life, fashion, advertising, communication media, show business etc. The use is less concealed and more public than in other types. For this type of user cocaine mythology, in which the drug is associated with elite, prestige and success, plays a more prominent role. This type consists mainly of professionals from the upper middle class, with medium qualifications. There are no (former) heroin addicts among them. They take cocaine intranasally. Their pattern of use shows a peak after which consumption stabilizes at a lower level. This type, has had no contact with the social-health network for drug problems. They have not committed criminal activities in relation to use. This type consists of 26 respondents, nine of them are women. The average age is 30 years.

Elitist type

"I only used a little bit at the beginning and now, for the last five years, I'm using it every day, even in the morning when I get up (...) it was little by little, but you get in deeper and deeper; your body gets more and more used to it until you find it's a real necessity." [B-EH2]

This type of user is characterised by certain elite backgrounds. They participate in closed circles that are marked by an exclusive lifestyle. A lifestyle in which prestige goods and activities (for instance expensive cars and playing golf) are seen as indicators of an elite status. Cocaine is regarded as one of these indicators and there is a more or less natural integration of cocaine within this lifestyle. It was in this context that the cocaine mythology of an elite drug was developed. The 'popularization' of cocaine use has meant, for this type, also its 'vulgarization'. Furthermore, characteristic for this type is the high availability and quality of the cocaine. All the respondents of this type belong to elite sections of society and are engaged in business. They have had problems caused by cocaine use and therefore have had contact with the social-health network for these problems. Criminal activities in relation to cocaine use are absent. In the sample there were three respondents of this type, all are men, and the average age is 35 years.

Commercial type

"When you spend so many years (selling), then you get it cheap on the side, from here, from there... It is the same as in industry. It costs me a lot of effort to get cocaine! I spend half the day getting it." [B-CT6]

For this type use is linked to dealing cocaine. Cocaine plays an important role as a form of life and business. Variable use, both in frequency and in quantity, must be kept under control, is characteristic for this type. Thus, these dealer-users cannot be a compulsive in their use because customers have to trust them and they must look after the economic viability of their activities. As cocaine dealers, the respondents of this type are engaged in illegal activities and their income is derived from this activity. They have had cocaine as their main drug throughout their lives. In the intensive period, they used daily in quantities from five to ten grams per month. The pattern of their consumption shows a peak and afterwards decreases. This type consists of 11 respondents. They are all men and the average age is 34 years.

Dysfunctional type

The characterization of this type follows Siegel's (1985) description: 'the compulsive use implies high-frequency and high-intensity levels of relatively long duration, producing some degree of psychological dependency (...). The compulsive patterns are associated with preoccupation with cocaine-seeking and cocaine-using behaviour to the relative exclusion of other behaviours (it

---

can mean breaking off all the relations; the only link is with cocaine). The motivation to continue compulsive levels of use was primarily related to a need to elicit the euphoria and stimulation of cocaine in the wake of increasing tolerance and incipient withdrawal-like effects'. In total, 18 respondents form this type. Two subtypes could be differentiated.

a. Pure cocaine addict

"I started using it at night; then in the morning; then I started using it to go to work... I just couldn't do without it and I was hooked. I gave up working because I just couldn't work, couldn't do anything at all. I didn't have time to work. I just had to move around in this little world because that's what attracted me." [B-DC2]

It is typical of this subtype that there is, or was, a compulsive use of cocaine but there has never been an addictive relation with heroin. Cocaine is the main drug and its use is not linked to specific occasions. Except for the initial period of use, cocaine is used on a daily basis. The main ways of use are snorting and free basing. There is no intravenous use in this subtype. As a result of using cocaine, they have had physical, psychological and financial problems. This made them seek help in the social-health network (more than half of them stopped using cocaine). Some respondents had engaged in criminal activities, mainly during the intensive period. Four of the seven respondents were men. The average age was 30 years.

b. Cocaine addict ex-heroin addict

"I was using cocaine intravenously right from the start. I didn't start by sniffing it or anything. I gave up heroin and was only using cocaine, taking a quarter and then half a gram and then later, well, there you are then, I was consuming everything I could get" [B-DP2]

Characteristic for this subtype is a compulsive use of cocaine that has replaced heroin as the main drug. Before using cocaine, they had taken heroin and other drugs. In general these drugs became less important after they started using cocaine. They can be considered former heroin junkies. The way of use is mainly intravenous. There is also very intensive free basing use and occasional nasal consumption. In some cases, the free basing use has substituted intravenous consumption. This substitution may be connected with the negative image of intravenous consumption (identified with heroin junkies). Throughout their whole career they used cocaine daily. Their pattern of consumption rises to a maximum point and afterwards decreases. During the intensive period and the current or last period, they consumed more than 15 grams per month. All of them have had physical, psychological and financial problems due to using cocaine. They were under treatment for cocaine, like they had been earlier (mainly for heroin). Criminal activities were engaged in during their addiction to heroin and in their intensive, and last, period of cocaine use. Although cocaine is the main drug, a functional use of heroin in order to counter the negative effects of cocaine can be seen. Five of the 11 respondents were women. The average age was 29 years.

Heroin addict type

"It's twice the ruin heroin is (...) You give yourself a jab of heroin and you can get by for five or six hours, but when you give yourself a cocaine jab, you need a new one the minute you take the needle out. Then, even when it's half the price of heroin... you need all you can get and more too... and you're just using more and more. And in the end, if you haven't got any heroin then God help you, because you could do heaven knows what.." [B-PI4]

The respondents of this type are all addicted to heroin. It is and has always been the most important drug. Cocaine is of lesser significance. Within this type, there is a functional use of cocaine to counteract the effects of heroin. The consumption of cocaine is intravenous and may, from time to time, be in very high quantities ('have a party'). Some of them consume cocaine and heroin together ('speedball'). They have a low level of education and all form part of the heroin sub-culture. They have carried out criminal activities in all the periods under consideration and their current income is derived from these activities. This type consisted of six respondents, five of them men. The average age was 28 years.

Former heroin addict type

"Now, if I consume more than a gram a day, which has happened sometimes, the truth is that then I begin to get worried. It's because everything is so dependent on the financial factor and also, of course, the relationship that you have with work." [B-EI2]

This type is characterized by a specific drug addiction life history. The consumers included in this type have in common a past of poly-drug addiction, especially their addiction to heroin, heavy cocaine consumption and, in general, having completed some rehabilitation treatment. During their itinerary a separation (rupture) has occurred in their previous relationships with the heroin sub-culture. Subsequently, in all cases, cocaine consumption is intranasal (sniffed) and in normal contexts; a kind of consumption that can be compared with that of the social, circumstantial and situational types. Before their first use of cocaine, in the Seventies, they have taken heroin and other drugs. They are ex-heroin addicts and have received treatment for this drug. Their pattern of consumption rises up to a peak and, afterwards, decreases until it stabilises at a lower level. They have been using cocaine for ten to 15 years and have had physical problems related to this. This type consisted of 13 respondents, eight of them women. The average age was 31 years.

6.2.2 Rotterdam

In the Rotterdam study, the criteria distinguishing the types are the place cocaine occupies in the life of a respondent (for example how much is he prepared to give up for cocaine) and the drug's influence on his lifestyle (for example whether using cocaine has affected friendships, etc.). In addition, the types are distinguished on the basis of differing reasons for use and functions of use. These distinguishing criteria can be brought together on two distinct but related general dimensions. The place and influence of cocaine can be classified under the dimension 'centrality of cocaine use in lifestyles'. Cocaine may have a peripheral, an intermediate or a central importance within lifestyles. Reasons for and functions of use can be classified under the dimension 'meaning of use'. This dimension is also divisible into three aspects. First, use can be characterized by subcultural aspects. It is linked to, and takes place within, a specific subculture, for example, in the artist scene users want to differentiate themselves from the common crowd. Secondly, use can be characterized by functional aspects; cocaine use has a clear function, e.g. to expand the world of experience. Thirdly, use can be characterized by the occasion. It is linked to certain, more or less specific, occasions such as Christmas and New Year's Eve. Each type is characterized by a specific combination of these two related dimensions.

The Burgundian type

"Between the dessert and the coffee with calvados the mirror is presented with a line of coke." [R-069]

Cocaine plays only a minor role in the lifestyle of this type. It forms a part of a lifestyle which is, above all, luxurious. Enjoyment is important but it remains within certain bounds. Cocaine is seen as a luxury leisure item to be compared with an exclusive dinner or bottle of excellent wine. Use is often consciously kept within bounds to avoid the loss of its exclusive character. It is seen in the same light as champagne or caviar. This does not, however, imply that cocaine enjoys the same status among, or is accepted in the same way by, other members of their circle. Their cocaine use is known only to a small group of close friends, most of whom take cocaine in a similar manner. It is not discussed openly in order not to endanger the social position of the user. The circuit of close friends (intimate circuit) plays a crucial role, both in obtaining and using cocaine. Discretion and trust are the key elements. The description of the Burgundian type is based on five respondents, one of them female. Their average age was 38 years, relatively old compared to the other types.

The experience type

"I am always experimenting with drugs and then at a certain moment I stand back and take a look at what I'm doing. How do I see things now and how did I experience it and what do I now think of that experience, and so on. Like that, I have been very consciously experimenting with drugs." [R-017]

Cocaine is of a limited significance for this type and is not a main activity. What they find important is gaining personal experience. They can assess the value of something only by trying it out themselves. It is in this context, that their use of cocaine must be considered. A range of drugs is tried out to see what effect they have. Cocaine is just one other in this range. Only after taking cocaine yourself, are you entitled to talk about it. In all phases of use, the positive and negative effects are weighed against each other. There is a sort of calculated use. In general, this means that use decreases or stops completely after a period of time, when it no longer provides a new experience and other things become more important. The experiment is rounded off and considered an enrichment of their life. The description of the experience type was based on six respondents, five of them men. Their average age was 27 years.

The situational type

"The end of the year at Christmas and New Year, that's when you take more. That's when there are all the feast days and you have more opportunity when you meet up with friends who are also users and have holidays and you certainly notice it. There's a big rush on New Year's Eve." [R-032]

This type is distinguished not so much by a specific lifestyle or predominant functional aspect which typifies use. The drug is of only fringe importance. A typical feature is the incidental use of cocaine, mostly linked to special occasions such as Christmas, New Year, birthday parties and other parties. Consumption is limited, either consciously or unconsciously. This can range from a more 'calvinistic' attitude to a more 'hedonistic' attitude. In the first case, various situations are regularly avoided if it is known beforehand that drugs will be used. In the second case, situations in which drugs are used are sought after and sometimes even created. Cocaine use does, however, remain linked to the special occasions. 23 respondents formed the basis for the description of the situational type, six of them female. The average age was 28 years.

The distinctive type

"I took it more because I find it great, you felt yourself so special. You feel you are one of the elect, very strong. And you were part of a certain group. That was really special." [R-005]

With this type, cocaine plays a particular role in the lifestyle. Use of cocaine is embedded in a subculture to which the respondent (temporarily) belongs. An important feature of the group is that the members are different to other people and distance themselves from the accepted norms of society. Drug consumption, including taking cocaine, provides an opportunity for them to oppose the more traditional values and norms of society which condemn drug use. It is also a means of acquiring a certain identity. The respondents belong to a select group of users. Many aspects of cocaine consumption (initiation, progress and any abstinence) are also largely determined by the group. Cocaine has played this role for some time in certain artist, squatter, and punk scenes. With the current widely spread use of cocaine at all levels of society, it would appear that this role no longer exists and this type of cocaine user is gradually disappearing. Six respondents formed the basis for the description of the distinctive type, five of them male. The average age was 33 years.

The hedonistic type

"I take cocaine because I want to get bonkers. Look, the whole week you sit there in your normal life and then it's the weekend and along comes a party. And then I have to take a lot of things to get yourself gaga." [R-064]

Sex and drugs and rock & roll could be the motto of this type. The central feature is unbridled pleasure seeking. They are looking for different kicks and trying to make life exciting and interesting. Anything which can increase enjoyment is permitted. They make wide use of the whole arsenal of pleasure seeking activities available in our modern society. This includes drugs. Cocaine fits ideally into this pattern and is considered the very height of pleasure. A number of short periods of intensive use is typical of this type. The hedonist does observe certain limits however. If drug consumption causes physical or financial problems which are at the cost of enjoyment, the respondents usually reduce their intake for a limited period. The description of the hedonistic type is based on six respondents, five males and one female. The average age was 23 years, relatively low compared to the other types.

The routine type

"I just find it nice. It's such an established pattern. (...) I mean, okay, you take it, you are only human and you need something." [R-074]

Cocaine plays a 'go-between' role and is not considered a luxury or exclusive drug. The main characteristic of this category of user is that cocaine has become more or less integrated into their life. It belongs to the daily activities in the same way, for example, as alcohol consumption and is not linked to special occasions. Christmas, New Year's Eve and other celebrations are merely additional occasions on which cocaine is taken. The use of cocaine is seen as normal and natural and is rarely queried. It has become a routine practice. These respondents, to a greater or lesser extent, deny that they experience problems caused by the regular cocaine consumption. If problems do occur, the respondent tends to limit his consumption for a short period. The description of this type is based on 14 respondents, three of them women. Their average age was 29 years.

The poly-drug type

"You are totally fixated on drug taking, you just put all sorts of other things aside, you don't have the peace and quiet to deal with them." [R-049]

The typical feature of this type is that both opiates and cocaine play a central role in their lives. These respondents belong mainly in the subcultural setting of the hard drug scene. A number of them have already developed a drug-centred lifestyle relating to a shorter or longer career of heroin use. For others, it was only at a later stage that heroin played a role. Within compulsive poly-drug use, cocaine plays the prominent role. Cocaine is considered by many to be more invasive and addictive than heroin. Consumption of cocaine is characterised by a broad range of problems. Physical, financial and psychological problems are often cited. Criminal acts undertaken to ease financial problems are not excluded. This involves subsequent contact with the police and lawcourts. Most of this category of cocaine user have had considerable experience with the social services. The description of the poly-drug type is based on 33 respondents, seven of them women. The average age was 30 years.

The cocainists

"You live only for coke." [R-028]

Cocaine plays a central and dominating role in their lives. The compulsive use of cocaine is typical of this category of user but, in contrast to the poly-drug type, heroin is excluded. There is little or no contact with the hard drug scene. Problems, including disrupted childhood years, are mentioned comparatively frequently as the reason for starting on cocaine. Over time, the compulsive use of cocaine has become a real problem. Physical, psychological and financial difficulties are frequently cited. These respondents see themselves as addicts. Some respondents are dealing in cocaine and these activities have involved them with the police and lawcourts. The description of this type is based on 13 respondents, all men. Their average age was 27 years.

6.2.3 Turin

As said above, the approach in Turin was different to that of Barcelona and Rotterdam. The analysis focused on styles of use throughout the different periods of use and had a more quantitative character, being supported by qualitative material. The classification according to the persistence or change of use styles (sequences of styles) made it possible to distinguish four classes of cocaine users[4]

Intimate group

This group can be illustrated by the following.

"It's a good thing with friends, it makes communication easier" [T-067].
"I use cocaine once, twice a month, a half to one gram. It's obtained by my boy-friend who knows the right people" [T-015].

These people maintain their intimate style of use over time. This means that they have never committed any crimes connected with cocaine and always have had legal sources of income. Cocaine is not used intravenously and always in a friendship circuit. A remarkable fact is that men and women are equally represented in this group. Most of the respondents (73%) have a job and more than three quarters are unmarried. Only a small proportion of this group (15%) experienced heroin before cocaine, but the majority had used cannabis (71%). Involvement in the cocaine market is relatively rare within this group (10%). A relatively large proportion (21%), on the other hand, obtained their cocaine mainly for free (21%). Ten percent of the users have had periods of abstinence that lasted longer than one year. Few respondents (8%) had had contact with either public or private treatment institutions, although a third of them had experienced some psychophysical problems in relation to the use of cocaine. The intimate user takes cocaine, maybe with alcohol, at parties, social gatherings, in his spare time, in a convivial atmosphere. After the first contact with the drug, use is kept within bounds. It apparently gives few problems and integrates well into normal life without any risk of totalization. It is mainly used to facilitate communication on social occasions. This group consists of 48 respondents.

Work group

Illustrative for this class is.

"It's like Super Goofy's peanuts, you take one when you can't make it otherwise." [T-077]

The definitions characterizing the intimate category apply to this group too. They have no illegal sources of income nor do they commit crimes related to cocaine use. They do not use cocaine intravenously and they consume it in friendship circles. To this, however, must be added that in some cases cocaine is also used in the work circuit and in others linked to work. The image of cocaine among these users, seems to be that of a tool to cope with work or to reach one's goal successfully. Four of the five respondents are men. In general their educational level is high and they are employed. Most are married. Two respondents have previously used heroin, but no one used it during the period they used cocaine. None had ever been involved in the cocaine market. Generally the use of these respondents can be regarded as continuous. This group consists of five respondents[5].

Hard group

The following quotations can be considered illustrative for this class of users.

"I used it every Sunday for going to the stadium with my group of hooligans. It helps to have more strength, to confront the opposed supporters." [T-058] *"I used cocaine, but smack (heroin) was still always the most important"* [T-074].

They keep the hard style of use throughout the three periods. This means that they always displayed criminal activities that they regarded as connected with use. Their sources of income are illegal. In addition, they use(d) cocaine intravenously or used heroin both during and before using cocaine. In some ways we can say that these users belong to a subculture presenting aspects peculiar to the underworld and aspects typical of the heroin addicts' world. The large majority (91%) of the respondents are men. More than half of them (64%) are unemployed. Generally they have a low educational level (74%). A remarkable feature is that almost three quarters (73%) started to use cocaine before the age of twenty and their length of use is less than five years. About one fifth of the users of this group have periods of abstinence longer than one year. Although 90% used heroin either before or during cocaine use, only one quarter of the respondents had contact with treatment centres. For some of them this was in prison, where they were subjected to compulsory treatment because of their heroin addiction and dependence on other substances. In that circumstance they agreed to be treated. The majority (74%), however, has no intention to stop using cocaine. This group is composed of 11 respondents.

Changing group

Illustrative quotations of these group of users are.

"I spent a period when I just overused. Now it follows cycles in relation to the possibility of getting it. I shoot it and snort it" [T-035] *"Occasions haven't changed, but the frequency has increased"* [T-040].

In contrast to the other three groups, they do not have a persistent style of use, but changed their styles over time. This is shown in table 6.1.

Table 6.1   Styles of use in each period in Turin (n = 32)

	first period	heaviest period	last period
Hard	5 (16%)	17 (53%)	7 (22%)
Intimate	18 (56%)	3 (9%)	8 (25%)
Work	2 (6%)	8 (25%)	11 (34%)
Others	7 (22%)	4 (13%)	6 (19%)

The majority (81%) of the group are men. In general they are employed and more than half of them (58%) have a high educational level. Three quarters of the respondents are unmarried. Initially more than a half of the subjects (58%) had an intimate style of use. In the heaviest period of use the hard style became most dominant (53%) while the work style became more important (25%). In the last or most recent period of use the distribution was more equal, although the work style of use is the most prominent in this period (34%). Several of the main characteristics of this group resemble the characteristics of the hard and work group. It is difficult to describe the multiplicity and varia-bility of their approach to cocaine. Although half of the respondents have used heroin before and 68% during cocaine use, the majority (64%) has never injec-ted cocaine. Three fifths reported having engaged in criminal activities. More than half of the users of this group (62%) reported a use of less than five years, and three fifths had not interrupted use for a period longer than one year. A majority (62%) of the changing users mention problems connected with cocaine use. These are mainly of a physical nature. However, over 80% have never contacted any treatment centres. A remarkable feature is that one quarter of respondents say that they want to cut down on use and almost three quarters have the intention to stop. This group is composed of 32 respondents.

6.3   Comparison

The above sections show a great variety in cocaine lifestyles and cocaine use styles. In this section a comparison is made of results of the qualitative analysis. The extensive qualitative material in Barcelona and Rotterdam made it possible to construct a typology of cocaine lifestyles. In Turin it was not possible to obtain the qualitative material to the same extent (partly due to the sociolegal context) and only a classification (preliminary typology construction) was possible. For this reason, the comparison will focus on Barcelona and Rotterdam. On the basis of the two typologies a more general typology was constructed. Within this general typology heroin forms the main emphasis: the different types can be grouped into two large categories, with the rejection of, or addiction to, heroin as the differentiating pivotal axis.

Most respondents were found in the first category, the non-heroin type. The social, circumstantial, situational, elitist, commercial and pure cocaine addict types of Barcelona, as well as the Burgundian, experience, situational, distinctive, hedonist, routine, and cocainists types of Rotterdam, fall within this category. Among these, we find an explicit rejection of heroin (although in some cases there was experimental use without developing addiction). On the basis of comparison, three types of cocaine lifestyles can be distinguished within this category:

• The leisure type.
• The instrumental type.
• The cocainist type.

In contrast, the respondents of the second category were or had been addicted to heroin. The cocaine addict ex-heroin addict and heroin addict types of Barcelona and the poly-drug type of Rotterdam fall within this category[6]. They socialize in circles using hard drugs, separated from the other types of users. In this category only one general type of cocaine lifestyle can be distinguished:

• The poly-drug type.

Summarizing, a total of four general types of cocaine lifestyles were distinguished. They are shown in diagram 6.2 and 6.3. Also an impression is given of the whereabouts of the initial types that were distinguished in Barcelona and Rotterdam. The (horizontal) position of the types in the diagram signifies the level (frequency and amount) of use which ranges from sporadic to compulsive use. In the remaining part of this section a description will be given of these four general types.

Diagram 6.2   No Heroin Category

	A - LEISURE TYPE	B - INSTRUMENTAL TYPE	C - COCAINIST TYPE
Level of use	(Sporadic .... Compulsive)		
Barcelona	Social Circumstantial	Situational Elitist Commercial	Pure cocaine addict
Rotterdam	Situational Burgundian Experience Routine	Distinctive Hedonist	Cocainists

Diagram 6.3   Heroin Category

D - POLY-DRUG TYPE

Level of use	(Marginal use of cocaine .... Central use of cocaine)
Barcelona	Heroin addict Cocaine addict ex-heroin addict
Rotterdam	Poly drug type

The leisure type

For this type, cocaine functions only on the periphery of their lifestyles. The consumption of cocaine is clearly linked to the context of leisure activities. They use together with friends and not alone. The ambience of use can be described in terms of affection, enjoyment, pleasure and celebration. The social relational aspect is an important characteristic of use.

"I find it a pleasurable social means of stimulation. A social drug." [R-041]

Cocaine is considered a nightlife drug. It is used at home with friends, but especially in bars, pubs and discotheques. Outside the leisure context, the use of cocaine is valued negatively and considered dangerous.

"Going out in the evening with cocaine? You're the 'king of the mambo' (...) you get to feel the first effects of cocaine the minute you get your hands on it" [B-SF25] *"No I take it only if things are going well. If I'm feeling rotten I simply don't take it."* [R-041] *"For me, I take it for having a good time. It's not the same as for the guy who uses it for working or getting hooked, because then it's a ruin (...) so long as it's used just for fun... ."* [B-SF47]

Often the use of cocaine is combined with alcohol and/or cannabis. Compared to these substances, cocaine is of minor importance. In general, use stays at a low level and sniffing is the only route of ingestion of cocaine.

"It is unusual to come across a party where there's no cocaine as yet another element like there being alcohol, like there being hashish fags or like music, or like sandwiches. In my circle, it would be most unusual to go to a party where there was no cocaine or hashish to smoke or alcohol ... Unthinkable, really." [B-SF15]

The lifestyle of this type of user can be described as integrated in society. They earn their income by legal means and have enough money to pay for all sorts of leisure activities and consumption. Criminal activities are taboo.

"A bit of luxury and, yes I can pay for that and that's the basis on which I sniff." [R-032]

Within this type, however, a distinction has to be made between two groups. One group is characterized by sporadic use. Cocaine is linked to very special occasions such as Christmas, New Years Eve, and special parties. The level of use always stays low. Any problems related to cocaine use are minor. From a problem perspective, this is a low risk group.

"It varies, a period of not taking it and then I start again. An average of once a month, one small line." [R-097] *"Among the people who take drugs for going out, no, nothing ever happens."* [B-SF52] *"I have never been a problem user. For me, it is really a social happening."* [R-003]

The second group is characterized by more habitual use, not so much limited to specific occasions. Cocaine also plays a role at the weekend when they go out. There are short periods in which use increases. During these periods minor problems can occur but if this happens the amount of cocaine used is quickly decreased. This type can be considered an intermediate risk group for cocaine-related problems.

"For the last six months it was in fact every week. Now I miss a week, once in a while. But the last couple of months it's in fact every week." [R-058] *"What I have noticed is that with the help of cocaine you can drink more. But I realise then that both these things are attacks on your body. You lose your appetite completely from the coke. But I keep a close eye on that."* [R-050]

The instrumental type

Within the lifestyle of this type the consumption of cocaine plays a more important role. It is not limited to the context of leisure. Other contexts, for instance work, become dominant. Cocaine is used not only together with friends but also with colleagues and sometimes alone. Not only social interactions, enjoyment, pleasure and celebration are seen as reasons or situations for use, but also work, sexual relations and to emphasise social distinction from other groups.

"I liked it best for working. When at times I'd be staying up all night writing, or making projects, or writing for magazines." [B-SC19] *"At the time it was sung about a lot. I still like rock music and in one way or other the impression has arisen that it belongs to that world, more than the other drugs. For me it was something tough-like, of 'Gosh, I play the guitar', and I felt something like 'now it's me playing too'. I was nineteen at the time."* [R-047]

For this type, the use of cocaine has a more instrumental character. This is also reflected in the locations of use. It takes place in more intimate locations, (at home with or without friends) and the workplace. The occasions where cocaine is used are not limited to celebrations and festivities during the night, but include, for instance, social activities related to work (dinner with colleagues).

In addition to cocaine, alcohol and cannabis, as well as other drugs like speed, LSD and XTC are tried, mostly on an experimental level. The patterns of this type show short periods of peaks in which the use can even be daily. The route of ingestion is not limited to sniffing. Other ways such as basing are also tried. Injecting is however out of the question.

"There are a lot of ways to use cocaine. With ammonia I like it best. I take a spoon, put in the cocaine and add ammonia. I heat it and continue heating it until it changes to oil and then I separate the oil, the base. I leave the cocaine to dry and become hard, and than it is only possible to smoke it. It is not possible to sniff, nor to inject, nor to do anything else with it, only smoke it. Then I smoke it. I have the glass pipe here and I use this to smoke (showed the pipe). I put it in and I smoke, no?" [B-CT6] *"I've seen it was a very destructive thing that was going to end up bad. From my experience too. I've seen friends, people who were strong on it, injecting it or with the base, using it daily, and I've seen how most of them ended up in the psychiatric hospital, dead or in prison."* [B-CT3] *"I never inject. I mean that I wouldn't even dare stick a needle in my arm. No, before you know it you are on heroin. I haven't had any experience with heroin."* [R-064]

This type of cocaine user has encountered a number of negative side-effects of cocaine use and they experience several problems in physical, psychological and financial sense. It is important to emphasise that there is a decrease of use after problems occur. It is even possible, for instance, that people break with their friends who keep on using. The impression is, that this type is asking for help, but does not know where to go to. The normal drug assistance is linked to junkies and therefore out of the question for this type.

"After a time like that you have a period that you are down and depressive. Particularly after a longer period like that it can be quite severe. I then stop for a while, with drinking too, a little while." [R-066] *"Three of the group have become real addicts, they are still taking it. The others have stopped in one way or another."* [R-051]

The lifestyle of this type can be considered as integrated in society, but to a lesser degree than that of the leisure type. Generally they obtain their income by legal means, but criminal activity is not excluded. For this type of user, cocaine has a specific meaning. Use has an instrumental character and is more dominant within the lifestyle. Regarding the more regular use and various routes of ingestion (especially basing), this group can be considered as a risk group, likely to incur problems.

The cocainist type

The lifestyle of this type is centred around cocaine. Use is linked to almost every context of their life.

"Everything else is subordinated to the coke." [R-082]

Generally cocaine is used with other cocaine users or alone. Use is not linked to special situations or occasions, but takes place at all possible situations and occasions. There are a multitude of reasons given for using cocaine, including problems (e.g. alcohol addiction, disrupted childhood). Use takes place in all kinds of locations: at home, in the workplace, in bars and so on. When these respondents go out, they look for specific locations where cocaine is used.

This type is characterized by a compulsive use of cocaine. Generally the frequency of use is daily and large amounts are taken. Several routes of ingestion are employed with sniffing and basing the most important. Injecting cocaine is strongly rejected, however. The use of this type can be described as dysfunctional and is related to many different problems. Physical, psychological, social and financial problems occurring at the same time is not unusual. When problems occur there is no decrease in use. Most of the users of this type are not able to decrease use without help. Many have (had) contacts with drug agencies or health services. Usually, these contacts were forced on the respondent either by his family, partner or by police/justice authorities after they had been arrested. If there are possibilities, they search for private help. The state-supported drug assistance serving poly drug users is avoided as much as possible, because they do not want to be associated with 'junkies'.

"I've only learnt the bad things of life, playing around, orgies, parties and all those things this brings. You've paid in exchange everything you got, health, family ... So it really doesn't compensate." [B-DC6] *"I spent a short period in a clinic at my own request. For alcohol and coke both together. I had then a very short period that everything got out of hand, that I couldn't control it any more. Then some very close friends of mine tried to talk to me about it. It didn't help to get off the coke. It was all far too short for that. The whole programme is something like two years."* [R-072]

The lifestyle of this type cannot be considered as integrated in society. A large part of respondents' income is obtained by illegal means. Often they have debts with dealers or they are themselves, to a greater or lesser extent, involved in the cocaine market. Criminal activities, as a way of obtaining money for cocaine, are not excluded. Cocaine is not seen as a social drug for pleasure for this type of users, it is something they are addicted to.

"I've stolen at home; I've taken their money, their jewels, their paintings..." [B-DC4] *"You get hooked in a stupid way, because I got hooked strongly on cocaine. You consume it and don't realize. You go out a lot, drink and... before you even realize it, you're using two or three grams a day (...) And this destroys you, it destroys everything."* [B-DC2] *"I have*

now reached the stage that I dare to admit: 'I am simply addicted to it, full stop'. It's crap to say that it is not the case if you are taking more than a gram a day." [R-072]

The poly-drug type

An important feature of the lifestyle of this type is the compulsive use of various drugs. Poly-drug consumption is the pivot of the user's daily life. They often use with other poly-drug users or alone. They use in all situations and for many reasons. Increasingly the reason given for taking drugs is the inability to do without it. The use of drugs is not linked to specific occasions. Their own home, that of other poly drug users, the street and dealer premises are the most important locations of use. They are part of the hard drug (heroin) subculture. Their poly-drug use is characterized by a high level of use. A large variety of methods of use are employed. Injecting, basing and chasing the dragon are the most important. Sniffing does not seem as popular among this type of users.

"You are always thinking of dope; if you go to bed and when you wake up." [R-045] *"First came heroin, one, two, sometimes three grams. Then came cocaine of 1, 2, 3, 4, 5, 6, 7, 8, 9, 10 to sometimes 15 grams in one day. We based. I got completely crazy from it but I was able to stop of my own accord at a certain moment."* [R-086]

In general the lifestyle of this group cannot be considered as integrated in society. Many people have no job and most have illegal sources of income. Delinquency related to the use of drugs is more or less common among this type of users. When looking at the role cocaine plays within the poly-drug use, a distinction has to be made within this type. On the one hand there is a group for which the use of cocaine plays only a minor role within the compulsive poly drug use. In other words cocaine is not central and heroin (and methadone) is more important. Heroin and cocaine are regularly used together ('speedball'). The problems they encounter are related mostly to heroin. Cocaine is only used when it is available. They refuse to turn to criminal activity in order obtain cocaine, but if cocaine were cheaper they would use more. From this point of view we can say that this group is a potential risk group for the use of crack, because (in first instance) crack is cheaper.

There is a second group for whom cocaine plays a major role. Although heroin and other drugs are used, they are of minor importance. The problems they encounter are mainly linked to the use of cocaine and are more severe than the problems linked to heroin. Most people in this group claim that cocaine is worse than heroin.

"Cocaine is a madness. At times I think that it's more dangerous than heroin is. Heroin destroys you physically but if you want to give it up, you can. Cocaine's a different matter. It gets your head hooked. It's very hard, much harder than you think, and until you fall to the bottom of the well you don't even see it." [B-DP10]

In cases in which cocaine is more important, there appears to be an increase of criminality and people are sometimes committing more violent acts to obtain cocaine.

"I was getting into it deeper and so the assaults had to be bigger for getting more money out of them" [B-DP2] *"Among the people that I've got to know, who are consuming cocaine strongly, it's all consumption and crime, consumption and crime. Twenty thousand, thirty thousand pesetas - whatever they can get! That's on an outlandish level, doing really stupid things, like at automatic cash machines at the bank and... I don't know what! (...) all under the effects of cocaine and with tremendous anxiety. It's a very dangerous thing"* [B-DP3]

For this group, cocaine appears to have a greater demand than crack. If they are not able to obtain cocaine, however, they might try crack as a second choice. Finally, it is important to mention that within this type there are different poly drug careers. Most of users of this type started with heroin and later on cocaine began to play a, more or less, central role. There are also users who first started with cocaine and in a later stage began using heroin (sometimes to counteract the negative effects of their compulsive cocaine use).

6.4   Conclusions and discussion

Cocaine is used by people in all social classes and by people with very different lifestyles. To fully understand the nature of cocaine use, it is important to consider the significance of cocaine within these lifestyles. The replies of the respondents show that the use of cocaine can have a wide range of meaning and significance in a user's life. Heroin, on the other hand, generally has a greater impact on the lifestyle of the user. Since cocaine is used regardless of social class, it is important to realise that the socio-economic status of the user does not play an important role in predicting whether he will use cocaine. Nor is this an indication of whether problems will occur due to the use of cocaine. In the case of cocaine, it is not so much the social background of the user nor the substance itself, but the meaning and significance attached to it that determines the occurrence of problems. The meaning and significance of cocaine are linked to the method and level of cocaine use. These, in turn, are strongly related to the problem aspects. Taking all this into account, a great variation of specific cocaine (life)styles were distinguished in each city. These can be grouped into two general categories in which the perception and use of heroin plays a major role.

First, there is a category of cocaine users, the majority, who do not use heroin. Within this category, there is considerable variation concerning the significance of cocaine use and the problems with use. Three types can be distinguished in this category. The leisure type is a group of users, well integrated into society, who use the drug only in context of leisure activities. In addition to cocaine, they also use legal drugs (alcohol, tobacco) or other illegal ones (cannabis). The problems derived from cocaine either do not exist, or are not prominent. The instrumental type is a group of users who, in addition to consuming cocaine in leisure situations, use it also as an energy booster at work or in other situations. Some users in this category experience problems (physical, psychological or financial). Usually, they manage to find their own solutions. Serious problems appear, in the long run, among a small minority of users who have no experience with heroin, the cocainist type. Cocaine has become the central element in the daily life of this type.

The second category of cocaine users also use heroin. Within this category only one general type can be distinguished: the poly-drug type. This user is part of the heroin subculture where the use of drugs is a central feature of the lifestyle. The consumption of cocaine generates serious problems, for the individual and for society. When cocaine, used intravenously or by way of basing, replaces heroin, the problems increase and become even more severe.

There is a clear distinction between those who use heroin in addition to cocaine and those who use only cocaine. These two categories appear to live in separate worlds. The dividing line between the two categories, is not only the use of heroin, but in particular, the acceptance or rejection of the 'needle'. The needle is seen as a symbol of social rejection, delinquency, AIDS and death. This perception is often shared by those who are addicted to heroin and see themselves as rejected by those who are not addicted. We can therefore identify method of use as a feature which distinguishes the categories: intranasal (no heroin) and intravenous (heroin). In this context, it is important to emphasize that basing as method of use is different from injecting. It does not have the same negative connotation and is not linked to junkies. However, basing is main method of use of most respondents in the non-heroin category who have encountered serious problems with the use of cocaine. Basing can be considered as dangerous as injecting because it can lead, in a relative short time, to compulsive consumption and problems related to this. In this context, the instrumental type and those users of the cocainist type who to date are only sniffing, may be considered a risk group because they lack a negative attitude towards basing.

To summarize, given that the intranasal route of self-administration of cocaine being the major route of ingestion, only a very small group has serious problems related to the use of cocaine (a small part of the cocainist type). The compulsive and problematic use of cocaine is more related to the consumers of heroin and the non-heroin users who have basing as main way of use. From an intervention point of view, it is important to take the division into two distinct groups into account. First there is a group of poly drug users, the vast majority of whom have problems with both heroin and cocaine. They can be reached more or less through, the existing channels of drug assistance. Secondly, there is a category of users not taking heroin, the vast majority of whom have no problems with cocaine. If problems occur they are solved by the user. The minority who do have serious problems will be reached only if the assistance offered is completely separate from that provided to the poly-drug category. Prevention, on the other hand, should be targeted to the risk groups such as the instrumental type for whom cocaine is related to a more habitual use and the risky basing way of use.

NOTES

1. It should be noted that this implies that the differences within the categories are as small as possible and the similarities as large as possible.
2. In the local report of Turin these are called 'patterns of use' (Merlo et al 1992). Here we use the concept of 'styles' since the term 'patterns of use' has elsewhere been used to refer to the level of use (frequency and amount) over time.
3. The analysis of 'classification with criterion' allows the researcher to identify the significant categories of each characteristic; in this case for each type of the typology.
4. The residual category 'Others' (4 subjects) will be left out of consideration.
5. While this group is numerically small, it appears to be paradigmatically relevant. Although only 5 interviews show these characteristics, more people encountered during the fieldwork gave hints in this direction. Unfortunately, they refused to be interviewed.
6. Due to the selected distinction pivotal axis heroin/non-heroin, the former heroin addict type of the Barcelona typology cannot be placed in the diagram. It is, however, important to mention this type of users when we look at the risks related to use. Presently, the consumers included in this type use cocaine intranasally and in normalized context. Nevertheless, they have a great tendency towards increased consumption and express the fear that they may 'lose control'.

CHAPTER 7

SPREAD, DISPERSION AND EXTENT

Aurelio Díaz, Uberto Moreggia, Marinus Spreen

In this chapter we compare the spread, dispersion and extent of cocaine use in the three cities. First a description of the opportunity structure will be given. After that we discuss the personal network analysis. In section 7.3 some specific analyses carried out in the three cities will be mentioned. The last section gives information on the extent of cocaine use.

7.1   Opportunity structure

We have already shown that the use of cocaine is widely spread in all three cities. It occurs among different age groups, very different job categories and in various places. As far as the places where cocaine is used and supplied, our study shows how cocaine presents many different aspects. Essentially, it's symbolic value is of a drug which enhances socializing. In this sense, most use occurs in places where friendship, leisure and entertainment factors are fundamental. We may therefore assert that cocaine is identified as a social drug. This is confirmed in our study which shows that the highest use occurs among friends and in entertainment circuits, and that it is hardly relevant in other situations. We have also observed that cocaine still upholds some of the myths in which its use is linked to social success, prestige, social promotion, modernity and snobbishness. On the other hand, cocaine is also used in the hard drug circuit. This circuit is totally separated from the circuits of socially integrated cocaine users.

7.1.1 Location

Cocaine use is particularly common in the night scene in certain pubs, bars and discotheques. In these places, use is often linked to alcohol consumption with, in many cases, cocaine facilitating a greater drinking capacity and 'making the night last longer'. Locations where cocaine is used are in different areas of the cities, in dispersed concentrations, usually frequented by a given type of clientele who visits specific pubs, bars or discotheques. Some of these places are linked to cocaine use and procurement, and even non-users are aware of this. In Barcelona, in particular, these are often identified as places where the modern avant garde and snobbish people meet. The most common setting is the evening entertainment scene. In this scene, generally, the use of cocaine is not public. The drug is not sniffed in front of everybody but rather in private areas on the premises, such as the toilet. However, increased control have made the users prefer to take the drug off the premises. In Barcelona and Rotterdam, the most common places are cars and parking lots. In Barcelona, automatic cash dispensers at banks, telephone boxes and other such places are also frequented. In Rotterdam, users sometimes go to benches in nearby parks.

"At first we used to sniff it in the toilets in the discos, but then this started to get dicey. The police started raiding the discos and they are always on the look-out. Sometimes, in some toilets, the security guard even comes in to check, whereas others don't let you in in twos and things like that. That's when we started using the automatic cash dispensers; they're very discreet and the surface is flat. Well, and of course the car. The most discreet place is inside your car; they can't say anything to you there. Go down to the parking lot of a discotheque and take a look for yourself; you'll find quite a lot of occupied cars." [B-SF44]

In any case, use in this entertainment circuit is essentially restricted to the privacy of a circle of friends and acquaintances and is dealt with discreetly; consumption is not shared with or mentioned to non-acquaintances.

"The people with whom I use cocaine I've known a very long time and I see them on an evening out." [R-041]

The preferred way of use is sniffing. The use of a mirror for preparing the cocaine lines is being replaced by any object with a smooth surface which is at hand, for example plastic documents, the palm of the hand, the fold between the index finger and thumb etc. The traditional snorting utensils (a hundred dollar note or silver tube) are now often replaced by a wide range of things such as pieces of straws like those used for drinking, empty ballpoint pens, rolled-up tickets, etc.

"If I go out I use a small mirror or I have to look out for something with a smooth surface. Like in the car, a cassette casing, it has to be a smooth casing. And then cut it up very fine. Make neat lines. And then with a tube, any kind - it doesn't matter - made of paper, a rolled up letter, anything, or if necessary just an empty pen casing." [R-001]

The high hazards and tight control of use in the city of Turin has led to the users sniffing the cocaine at home before going out, with use in public places being rare. In Barcelona and Rotterdam, too, there is preference for private use before going out, in order to avoid dangerous situations.

Social events

No use of cocaine was noted at social events occurring during the day. At such events, use is limited mainly to hashish and to alcohol. Some evening social events, particularly those that last until morning, are more suitable occasions for using cocaine. In Barcelona, one important party where cocaine use is quite common is that of Saint John's Eve, June 24th. The traditional parties (Festes de Sant Joan) held that night, in private or public, last until early morning, thus favouring the use of cocaine, and high consumption of alcohol. The term 'Coca de Sant Joan', normally referring to special cakes baked for this occasion, has a specific meaning for cocaine users. In Rotterdam, Christmas and New Years Eve are important events for cocaine use. For users 'a white Christmas' refers to a very special kind of snow.

Private meetings and parties

Cocaine use in private circuits, at home, is important in all the three cities. It is the most important location where cocaine is used in Turin. In Barcelona, on the other hand, consumption in private circuits is less important than in the entertainment circuit. Those who use the drug on festive occasions often do so only in public, and use at home is sporadic. Cocaine is nearly always used in the evening or at night with friends and in a private and welcoming atmosphere.

"At times at someone's home, with friends if we are all there, looking at movies or just talking... but on the whole when we go out to the discotheques and bars." [B-SF34] *"We used to have a circle of friends where everybody snorted. Well, some of them did, others didn't. At times, we all gathered at someone's place for snorting. It was fun. There always was someone who had some."* [T-070]

At times, it can also be used at a private home in a separate area so as not to disturb non-users, to restrict access or for other reasons.

"At parties with close friends where everyone likes cocaine it's used openly. At larger parties there is a certain discretion. We withdraw to another room, not secretly just discreetly. I don't openly smoke hash in public places, or take it at formal receptions. Same sort of discretion." [R-100] *"I've never taken coke on my own. It's a social happening. I've sometimes finished a leftover, but that's all."* [R-060]

The right occasion may coincide with celebrations (Christmas, New Year, anniversaries, weddings, etc.) but the drug is also used when friends get together with no special reason, when pleasure and entertainment are sought. The higher the level of use of cocaine in the individual's lifestyle, the more private it becomes. The circles in which they participate become smaller and within these circles the procurement and use of cocaine become more important. When a user cuts back on his intake, he usually abandons the user circle.

"Those days, circles were formed around (drug) use and all of us were involved in high use levels; the relationship established through cocaine, for as long as it lasted, was excellent, very good and very deep, everything you fancy. But, I believe that when you stop using it, the whole relationship falls apart, too." [B-EH3]

When use reaches the point of being compulsive and problematic, the relationship with the circle of users is often severed and the user takes the drug on his own.

"Drugs set you apart, they leave you out on your own. We were a large group, 10 or 15 people, they were all on the same wavelength (...) but then you start isolating yourself, and then the day comes when you don't go out drinking any more and start to use it on your own." [B-DC3]

Workplace

Cocaine use at work is less significant in the three cities. Use at work usually occurs when the profession or work activity is carried out in settings where the use of cocaine is open or semi-open (the night world, art, fashion, publicity, show business, etc.). In these cases cocaine is easy to obtain and, therefore, likely to be used. Sometimes, cocaine is even sold where the work is carried out. The reason for use may be as an energy-booster for long working hours, due to stress at work, or for certain periods of time that require a concentrated effort. We see how, for example in Barcelona among the situational type, periods of maximum use coincide with periods of use at work.

"When I had to go back to my job, I took it with me. Then I went to powder my nose and then I could keep going." [R-050].

Hobby/sports

It is only in Rotterdam we encountered cases of individuals who consume the drug in these circles. For example, amateur musicians who use cocaine before or during a performance or rehearsal with friends. In Barcelona and Turin, no respondents were found who use in these environments. This difference between the cities might be due to the special attention Rotterdam paid to this circuit during the field work.

The street

The street, as a location of use, is exclusively linked to the hard drug scene. There is a clear division, in all three cities, between the use of cocaine in certain socially integrated population sectors and its use in heroin user circles. The latter's negative image also covers the way of use (intravenous or basing, although in Barcelona and Rotterdam basing occurs also outside the hard drug scene). The effects of cocaine use in the subculture of heroin are the most serious, not only with regard to individual problems but also to consequences on the community level. In Rotterdam, this situation can be observed near the central station where there is a wide range of opiates and cocaine preparations available (white cocaine, cooked coke). Furthermore, heroin addicts in Rotterdam use drugs in dealer's premises and sometimes they stay in private locations (home) to use alone or with one or two other heroin addicts.

"At the dealer's place I usually wasn't allowed to shoot. Then I went in the woods, to a squat or home, anywhere where I could get water you could say." [R-059].

In Barcelona, too, cocaine use can be found on the streets in some areas in outlying districts linked to the heroin problems. In Barcelona, all the respondents consider the cocaine world to be incompatible with heroin subculture. Heroin and its users (addicts) are identified as a closed world of outsiders, a problems sector. Users make a clear distinction between this world and their own (cocaine) world.

"(The cocaine one) is a more gentlemanly one than the heroin one (...) the other (that of heroin) is the lowest you can get." [B-DC6]

This situation is even more acute and radical due to the AIDS problem. The identification of two incompatible and opposed worlds, as well as that of the street heroin against private, sniffed cocaine, applies equally to the three cities.

7.1.2 Cocaine supply

Access to cocaine is easy for most respondents in all three cities. In spite of changes in the laws relating to drug use in Italy, there appear to be no changes in the way users obtain cocaine in Turin. When use begins, the most common way of procuring cocaine is through invitations at a party or from acquaintances. After that, the newly initiated user gets in touch with other users who later provide the connections for procurement. Increase in use often involves closer connections with higher rungs of the sales and distribution chain, coming over time nearer to the source.

The different levels in the distribution chain have specific features of their own concerning the relationships generated by the transactions in cocaine. At some levels where involvement is less, relationships are based on friendship and the actual transaction is of secondary importance. When closer ties are established with the market, the relationship with distributors is based on knowledge and trust, thus assuring a discreet interchange, a good quality product and more favourable price. The cocaine market, above all at the middle and lower levels, appears to have a structure and dynamics of its own. In this sense, it is separate from the heroin market and that of other drugs, in spite of the fact that one can at times find cocaine with others drugs (hashish, speed, XTC, etc.). Although cocaine is present on the heroin market, users approaching these sellers are those who also purchase heroin. Below, we give an overview of sources and places of procurement.

Procurement by invitation

The habit of inviting people to use cocaine is widespread, above all inviting those who have never tried it or whose use level is very low. Frequently, at a party or social gathering, users take advantage of the situation in order to obtain the drug for free.

"I'm one of those who takes advantage of others, those who never have any and always use it. Well, I don't know, it depends on the people's generosity... The people who invite you are different. That's the circle I usually move in, which can be wider or narrower. At times I've used it with people who aren't regulars in my closest circle but who belong to it through friends and friends of friends..." [B-SF42]

The habit of inviting people is similar to buying a round of drinks or the invitation to share a packet of cigarettes where a lot of users, particularly women, are offered something for free but do not often buy a round in return. In Barcelona and Rotterdam, we came across users with a low rate of use, who obtain cocaine almost exclusively by invitation when mixing with friends or circles of users who have a higher rate of use (from the leisure type; social type in Barcelona and situational type in Rotterdam).

"I always had it offered by a friend or a partner. I've never bought it myself since I consider it's a waste of my money. Then I thought I wouldn't bother with sniffing and I felt fine then, too. I've never had the feeling that I just had to get it." [R-093]

In Barcelona and Rotterdam, invited consumption occurs commonly in all types of users, except among heroin users who purchase it all for themselves right from the beginning. In addition to among the leisure type, the frequency of invitations is especially high among dealers who often use it on the invitation of their clientele.

"They're offering it to you because it was in the theatre atmosphere and there were people using it and they always invited you, and then later on I started to buy it myself and later on, now, well I buy it so as to return the invitations. That's typical in these settings... It's like a society rule. If so and so has it, well you do to, and everybody makes their contribution." [B-SC22]

Sometimes, offering cocaine is used as a way of demonstrating high social status or to as a means to obtain more status. In Turin, for some (ex-)heroin users it is a way to be involved in a social ambience which is otherwise difficult to join.

"In the well-off circuits coke is offered as if it were good wine. Some heroin addicts offer coke as a social advancement, for a promotion in status, category, in order not to be considered addicts." [T-086]

Traffic between friends

The procurement of cocaine within the circle of friends is seen in all three cities. The group of friends contributes a given sum of money and one of them buys it for all. The friend buying it is the one who has the contact with the dealer. In some groups it is always the same person, in others it rotates. Sometimes the profits for the friend-distributor are a supplement to, or all, the cocaine he himself uses ('he who divides... keeps the best part for himself'). In Barcelona we saw that the group of friends who purchase cocaine together had often earlier purchased illegal drugs (cannabis etc.) as in a group. The purchase and distribution of cocaine does not follow the pattern of other drugs.

"The type of distribution (that of cocaine) is very different from that of other types of drugs. It's much more (likely to be) from a friend who knows three or four other friends and who gets it from one and who has three of four to whom he distributes it... Cocaine peddling is very split up and also it's a very.... closed circle, one based on contacts with friends." [B-SF33]

A third of the respondents in Barcelona and a quarter in Rotterdam had always purchased cocaine without going outside the group of friends. Nearly all of them are low level users and of the leisure type. In Turin in 43% of the cases, cocaine was obtained mostly through friends and the procurement of cocaine through friends may even have links with the production centres. A friend may go to South America and import considerable amounts of cocaine which is exclusively sold to a small group of friends. With this system, both quality and low cost are assured. In Barcelona and Rotterdam, there was no evidence of such close links with the market when buying between friends. Procurement through friends is essentially among low and medium level consumers. Some of them obtain cocaine exclusively by invitation, others also buy the drug at public premises or from house dealers.

Nearly all the users who deal among friends already have a high rate of use themselves and in this way they are able to sustain this level or increase it without having to reduce their other expenses (circumstantial and situational types in Barcelona and routine type in Rotterdam).

"I buy it for lots of friends, too. (...) They come to me to get it. It's a sort of deal between friends." [R-024]

The supply between friends has a clear significance related to the nature of use: social relationships, discretion and mutual trust. But it also has other functions such as: avoiding the risk involved in getting in touch direct with the market; involving a single person in the dangerous relationship and paying him in cocaine for his own use; and improving the quantity/quality/price ratio and avoiding fraud.

Dealers

An increase in the level of use generally results in a more direct relationship with the market in spite of the fact that, at times, users continue to purchase among friends. Often it is the friend-distributor who puts the person in touch with the dealer, and he in turn becomes a dealer upon increasing his friendships and acquaintances who will later form his clientele. The dealer can sell at his own home or take the goods to the customer's home or workplace. Contact by telephone (particularly in Barcelona) is important. Usually the contact is between people who already know each other. The approach is always through someone who introduces a friend, followed by direct contact. In this type of relationship, transaction and payment are the main link but often friendships arise which assure mutual trust and discretion and avoid the risk of conflicts. This inter-dependent link avoids risks and intermediaries, assuring that the cocaine obtained is of better quality and/or price. The purchase and sale relationship is always a private one based on prior knowledge and trust. It never takes place on the street.

"It's always been a very intimate circle, with friends, acquaintances, in houses. The dealer was always very respectable. Finding good cocaine is not easy; as with heroin, you've got to get into circuits that are usually quite close." [T-071] *"If you go to the dealer directly, this means you're going through fewer hands (...) he is also in control and will say this is better or this is worse and so on. Maybe he'll weigh it on the scales in front of you and say, 'look, here's an exact gram'."* [B-SC15]

These dealers never sell heroin. They have no connections with the opiates market, in spite of the fact that from time to time they distribute other illegal drugs (hashish, speed, XTC, etc.). Dealers who are not linked to the subculture of heroin also avoid the type of clientele that also uses heroin.

"Only coke (...) the dealers I know have a sort of code of honour and sell only coke. It's the rule among drugs. If you want to be respected you sell only coke." [R-022]

7.1.3 Settings of supply

Often, dealers sell on the premises where the drug is used (in pubs, bars or discotheques)[1]. This form of procurement appears to occur more frequently in Barcelona although in Rotterdam, too, there are premises where exclusively cocaine is sold. Dealers may sell on the premises, usually with the owner's knowledge, the owner or his staff are sometimes the ones selling the drug.

"We got friendly with the staff at the disco and there, in discos, it's always easy. Anyway, all night workers move on from one thing to another even though they're not recognized." [B-CO22]

In Rotterdam, employees of these premises sometimes are the people who distribute the drug and they may also sell other drugs such as hashish, speed or XTC. In some coffee shops in Rotterdam, you can get cocaine as well as soft drugs, but most of them avoid selling heroin so as to keep hard drug users away, for fear of contamination. In Turin, the sale of cocaine in public places is much less common than sale at home.

Purchasing cocaine on the street market is largely restricted to heroin users. The areas and places where it is on sale are previously known and identified with the sale and use of hard drugs. In Rotterdam, street prostitutes sometimes act as couriers of cocaine and receive part of the drug purchased as payment. Cocaine is also found on the heroin market in all the three cities. The users know the specific areas and spots in the different districts where they can find it, and even adapt to the time of day or night (as from early morning, only at certain places). The cocaine market in this circle of heroin addicts has a style of its own and has no connection with the other market in which exclusively good image users buy it. Those that have not been moving in these circles before do not go to this market.

7.2   Analysis of personal networks

The degree to which a respondent nominates other users in the various circuits provides a picture of his relationships with other cocaine users. In Rotterdam, the respondents nominated other users whom they classified as follows: 38% in the entertainment circuit (472 users), 17% at the workplace (209), 25% in the intimate circuit (310), 6% in the hobby/sports circuit (79) and 14% in the hard drug scene (169). In Barcelona and Turin, the same classification was used at the beginning but later on several circuits having the same importance were considered as separate categories. For this reason, in order to facilitate the comparison, we are only presenting the proportion in each circuit. In Turin, the results were as follows: the entertainment circuit 34%, workplace 15%, the intimate circuit 46%, the hobby/sports circuit 1% and the hard drug scene 3%. Barcelona shows the following results: the entertainment circuit 49%, workplace 10%, the intimate circuit 30%, and the hard drug scene 12%.

While there are many similarities between the three cities, there are also certain differences. Particularly noticeable is the higher proportion of users at the workplace in Rotterdam and Turin, the predominance of the intimate circuit in Turin and that of the entertainment circuit in Barcelona. In Rotterdam, an analysis of the average number of nominees per circuit shows that the respondents know users both in their own circuit and in other circuits. This is an indication of the open nature of cocaine use. Barcelona showed a similar pattern, but less marked (see tables 7.1 and 7.2).

Table 7.1 shows that the respondents from the entertainment circuit and the intimate circuit know roughly the same number of users. These respondents mention more or less the same number of other users in the five nominees circuits (there is no significant statistical difference). It is remarkable how few other users they mention in the hard drugs scene, showing that the connection with this circuit is very weak. The difference between the respondents from the entertainment circuit and intimate circuit concerning the average known number of other users in the entertainment circuit is also noteworthy (6.0 and 4.1 respectively). This difference can be explained by a so-called snowball effect which is discussed in detail in appendix C.

Table 7.1   Average number of nominees per circuit in Rotterdam (standard deviation between brackets)[2]

	Circuit nominees				
	Entertain-ment circuit	Work place	Intimate circuit	Hobby/ sports	Hard drug scene
Circuit respondent					
Entertainment	6.0	2.0	4.1	0.8	0.9
n=34	(3.9)	(3.0)	(3.3)	(1.7)	(2.3)
Intimate	4.0	2.2	4.1	0.9	0.9
n=34	(3.7)	(3.0)	(4.0)	(1.9)	(1.8)
Hard drug	3.0	2.2	1.8	0.6	4.4
n=22	(3.3)	(2.9)	(2.2)	(1.3)	(4.1)

In Barcelona, there are some significant differences between the average number of persons to whom reference is made (table 7.2). A remarkably low percentage of persons is nominated in the intimate circuit. This is due to the fact that a certain kind of cloning of the respondents exists among nearly all those referred to pertaining to this circuit: if the respondents own use does not take place in the intimate circuit, he mentions hardly anyone in that particular circuit. The respondents in the intimate circuit nominate consumers from the same circuit and from the entertainment circuit. On the other hand, those in the entertainment circuit nominate, in particular, people in this same circuit (with a high level of endogamy). Compared to Rotterdam, few respondents in the entertainment circuit nominate people in the intimate circuit. In Rotterdam, users in the hard drug scene systematically nominate different numbers of users in the five circuits while this is not the case with respondents in the entertainment and intimate circuits. Respondents in the hard drug scene mostly nominate users in the hard drug scene and, to a lesser extent, in the entertainment and intimate circuits. In Barcelona, the low number of cases in this circuit made analysis impossible.

Table 7.2   Average number of nominees per circuit in Barcelona

	Circuit nominees				
	Entertain-ment	Intimate	Hard drug	Intimate and entertain-ment	Intimate, entertain-ment and work
Circuit respondent					
Entertainment (n=39)	4.9	0.5	0.2	0.4	0.4
Intimate (n=17)	1.7	1.9	0.0	0.5	0.5
Hard drug (n=3)	0.7	0.0	0.0	0.3	1.0
Intimate and entertainment (n=38)	1.0	0.2	0.2	2.3	1.4
Intimate, entertainment and work (n=29)	0.3	0.4	0.0	0.8	2.3

In Turin, in order to calculate the average number of nominees per circuit, it was decided to consider the entertainment circuit and the intimate circuit together, distinguishing them from the work circuit and the hard-drug circuit (table 7.3).

Table 7.3 Average number of nominees per circuit in Turin

	Circuit nominees		
	Entertainment/Intimate	Work	Hard drug
Circuit respondents			
Entertainment/Intimate (n=26)	1.6	0.6	0.1
Work (n=15)	1.9	1.9	0.1
Hard drug (n=6)	2.5	0.3	0.1

A global view of table 7.3 shows that the respondents nominated especially persons who were encountered in similar circuits (entertainment/intimate and work) but that there is a low presence of subjects in the hard-drug circuit. That holds also for the respondents in the hard-drug circuit, and shows considerable differences between the situation pertaining in Turin and that in Rotterdam. The analysis carried out in Turin is the only one showing the spread of cocaine in the respondents' personal networks. On average the networks are composed of 58% of persons who use or have used cocaine and of 42% non-users.

Cocaine contacts and mutual knowledge of cocaine use

The degree of mutual awareness (between nominees and respondents) about the use of cocaine presents significant differences in all three cities, particularly in Turin. In Barcelona, the proportion is around 97%; in Rotterdam it is slightly lower (93%), whereas in Turin it is 70%. The analysis of the entire personal network in Turin shows that the degree of knowledge about the respondent's use among the non-users nominated by him is very low (17%). This figure, linked to the above one, suggests that cocaine use rarely occurs in the presence of persons who do not use it. This points to interpreting cocaine use as a social phenomenon, but a concealed one. Qualitative information from Rotterdam and Barcelona confirm this aspect of the phenomenon, though with a lesser degree of concealment (see section 7.1.1).

The degree of significance of cocaine in relationships between respondents and nominees provides information on how people behave towards cocaine. In order to determine the significance of cocaine in the relationship between a respondent and a nominee, a scale has been constructed in Rotterdam, also applied in Barcelona (diagram 7.1).

Diagram 7.1   Degree of 'significance of cocaine for a relationship'

Degree of significance	Description of scale value
0	The contact never takes places in a context of cocaine and the nominee does not know if respondent uses (has used) cocaine.
1	The contact never takes places in a context of cocaine but the nominee knows that respondent uses (has used) cocaine.
2	The contact sometimes takes places in a context of cocaine and the nominee knows that respondent uses (has used) cocaine.
3	The contact usually takes places in a context of cocaine and the nominee knows that respondent uses (has used) cocaine.
4	The contact always takes places in a context of cocaine and the nominee knows that respondent uses (has used) cocaine.

The scale is basd on the following idea. Each respondent nominates other users who he knows to be taking (or who have taken) cocaine. If a respondent says that the contact with the person he nominates was never in the context of cocaine, then cocaine is not important in the relationship. If, on the other hand, a respondent nominates someone he always meets in the context of cocaine and there is mutual knowledge of cocaine use, then cocaine has a significant place in this relationship. Diagram 7.1 shows the different values of this scale in Rotterdam and their significance. In Barcelona the scale is similar, but it includes an intermediate level of contact between sometimes and usually; for this reason, the scale ranges from zero to five.

In table 7.4 we show the results in Barcelona and Rotterdam. What is most striking are the similarities between the two cities. From this table it appears that half of the contacts nominated by the respondents in Barcelona and Rotterdam are only in the context of cocaine, and that the users know of the mutual use of cocaine (value 2). In both cities there is also a similar percentage of contacts that never take place in the context of cocaine although use is mutually known. If we consider these results (scale value 1 + 2 = 65%, 66%), then, apparently, the role of cocaine is not very significant for a large number of the nominated contacts. This is a further indication that the use of cocaine within the sample population is of an open nature. People know about each other when they take cocaine, but they meet mainly in situations which have no connections with the drug. This same conclusion is valid for the situation in Turin.

Table 7.4   Spread of values of the scale 'degree of significance in a relationship'

| | Rotterdam | | Barcelona | |
	n	%	n	%
Scale value				
0	59	5	37	6
1	156	13	86	14
2	643	52	319	52
3	169	14	77	13
4	197	16	47	8
5			44	7
Unknown	15	--		
Total	1239	100	610	100

In table 7.5 the frequency of contacts between nominees and respondents in a context of cocaine use in the three cities can be observed. This table, also, shows a lot of similarities between the three cities. Apart from those we have already mentioned, the existence of relationships in which the use of cocaine plays a relevant role should also be underlined. In Rotterdam and Turin respectively, 16 and 18% of the contacts take place in the context of use; in Barcelona this proportion is lower (8%). These relationships take place on the whole among users from the hard drug scene and/or among those with a high level of use.

Table 7.5   Frequency of contact between respondents and nominees in a context of cocaine use

	Barcelona		Rotterdam		Turin	
	n	%	n	%	n	%
Always	48	8	198	16	26	19
Usually	88	15	171	14	13	9
Sometimes	371	61	650	53	71	51
Never	96	16	205	17	30	21
Missing	7	1	15	1	10	7
Total	610	100	1239	100	150	100

Length of the relationship

As may be seen in table 7.6, the relationship between the respondents and the nominees lasts a long time, about five years on average (higher in Turin and lower in Barcelona) with no significant differences at all between the cities.

Table 7.6   Length of the relationship between respondents and nominees

	Barcelona		Rotterdam		Turin	
	n	%	n	%	n	%
1 - 6 months	25	4	21	5	4	3
1/2 - 1 year	58	10	30	7	5	4
1 - 2 years	79	13	56	13	15	13
2 - 4 years	137	23	108	24	27	23
4 - 7 years	128	21	90	21	40	34
More than 7 years	180	30	135	31	28	23
Total	607	100	440	100	119	100

Joint use between respondents and nominees

As shown in table 7.7, there is a predominance of respondents in Turin and Barcelona who say the have used cocaine with the nominees during the last six months. In Rotterdam this proportion is lower (less than half). It looks as if the use of cocaine in Rotterdam is more accepted and more open than in Barcelona and Turin. In spite of that, and we said earlier, cocaine does not usually play a major role in the relationship.

Table 7.7   Joint consumption between respondents and nominees during the last six months

	Barcelona		Rotterdam		Turin	
	n	%	n	%	n	%
Yes	393	65	202	46	94	67
No	186	31	212	49	27	19
Not applicable	22	4	22	5	19	14
Total	601	100	436	100	140	100

The location of joint use (see table 7.8) presents some slight differences. There is a predominance of private places and entertainment circles, slightly more in Barcelona (93%) and Turin (90%) than in Rotterdam (84%). In Rotterdam, use in the hard drug scene is higher (9%) than in Barcelona (2%) and Turin (3%). Apart from these differences it is remarkable that the figures show the same trend. Probably, the high proportion of use in private locations is an indicator of the trend towards a recent increase of use in the private sphere, as already mentioned.

Table 7.8   Location of joint consumption between respondents and nominees during the last six months

	Barcelona		Rotterdam		Turin	
	n	%	n	%	n	%
Private	206	40	69	36	56	38
Work	25	5	12	6	10	7
Entertainment	269	53	93	48	76	52
Hobby/sports	--	--	2	1	--	--
Hard drugs	9	2	17	9	4	3
Total	509*	100	193	100	146*	100

* In Barcelona and Turin the respondents could mention more than one location.

In closing, a limitation of the above analysis needs to be mentioned. It is difficult to decide to what degree the similarities and differences between some of the results in the three cities may be due to differences in the methodology, to the emphasis given to one aspect or another, to the nature of the phenomenon under study or to its setting in terms of values and social practices (social-cultural variants, the existence of given cultural patterns in drug use, juridical framework, etc.). This limitation, however, is common throughout the social sciences and the source of their epistemological uncertainty. In studies of hidden populations this indeterminacy is especially apparent.

7.3 Some particular network analyses

7.3.1 The role of cocaine in relationships in Rotterdam

In order to obtain another description of the spread of cocaine in Rotterdam, we also investigated the role procurement cocaine plays in the relationships. There are a number of ways in which procurement of cocaine can occur in a relationship. There could be a symmetric relationship: the respondent and the nominee purchase the cocaine in turn or together from a third person. There might be a one-sided procurement of cocaine (asymmetric relationship) which means that the respondent buys or obtains the cocaine from the nominee or vice versa. Finally, there is the possibility that a relationship exists which is unrelated to obtaining cocaine.

The frequency with which people meet in a context of cocaine and the manner in which cocaine is obtained in a relationship provide information on the extent to which the relationship is influenced by cocaine. If the contact is always, or usually, in the context of cocaine and if the respondent buys or obtains cocaine from the nominee or vice versa (asymmetric relationship), this relationship is to a large degree influenced by cocaine. If the contact always takes place in the context of cocaine but cocaine is obtained or bought in turn or jointly through a third person (symmetric relationship) the relationship is to a large extent influenced by cocaine. This is however of a less instrumental nature since the relationship is symmetric. Diagram 7.2 shows the different combinations of the scale 'extent to which a relationship is influenced by cocaine'[3]. Every combination gets a value. Zero means that the relationship is not at all influenced by cocaine. Six means that the relationship is strongly influenced by the procurement of cocaine.

Diagram 7.2   Scale of 'extent to which a relationship is influenced by cocaine'

Degree of significance	Description of scale value
0	The contact never takes places in a context of cocaine and there is no connection relating to the procurement of cocaine in the last six months. The nominee does, however, know that the respondent uses cocaine.
1	The contact sometimes takes places in a context of cocaine but there is no connection relating to the procurement of cocaine in the last six months. The nominee knows that the respondent uses cocaine.
2	The contact sometimes takes places in a context of cocaine but there is absolutely no connection with the procurement of cocaine. The nominee knows that the respondent uses cocaine.
3	The contact sometimes takes places in a context of cocaine and the nominee knows that respondent uses cocaine. They take turns to obtain cocaine or buy it jointly from a third person.
4	The contact sometimes takes places in a context of cocaine and the nominee knows that respondent uses cocaine. In contrast to scale value 3, cocaine is obtained one-sided. When there is contact in the context of cocaine, either the nominee supplies the respondent or the respondent supplies the nominee in these relationships.
5	The contact usually/always takes places in a context of cocaine and the nominee knows that respondent uses cocaine. They take turns to procure cocaine or buy it jointly from a third person.
6	The contact usually/always takes places in a context of cocaine and the nominee knows that respondent uses cocaine. Cocaine is obtained one-sided so that the position of cocaine has an instrumental nature in the relationship.

In table 7.9 are given the scale values of 259 contacts and 63 respondents[4]. It appears that nearly half the contacts (scale value 0 + 1 + 2 = 48%) of the last six months between the respondent and the nominee had nothing to do with the procurement of cocaine, regardless of whether they meet in the context of cocaine never, sometimes, or always. Nevertheless, these people are aware of each other's cocaine use. This is once again an indication of the open nature of cocaine use in the sample population. Nearly a quarter of the contacts occur occasionally in the context of cocaine, where the cocaine is jointly obtained from a third person or where the respondent and the nominee take turns to procure it (scale value 3). More than one tenth of the contacts occur occasionally in the context of cocaine, just as in scale value 3. The difference is, however, that in these relationships the cocaine is supplied by one person (scale value 4). Finally, one sixth of the contacts consists of contacts which always take place in the context of cocaine and in which cocaine is obtained either jointly through a third person or in turn (scale value 5) or in which one of the two supplies cocaine to the other (scale value 6).

Table 7.9   Spread of scale of 'extent to which a relationship is influenced by cocaine'

scale value percentage	number of relations	
	n	%
0	29	11
1	80	31
2	16	6
3	63	24
4	30	12
5	21	8
6	20	8
Total	259	100

With the help of a two-level regression analysis we made an exploratory search for the factors which have a major influence on the role of cocaine in the relationships[5]. The reason for using a two-level regression analysis is that units from two nested levels are included in the analysis.

Table 7.10   Factors of influence on the role of cocaine in relationships

	Regression coefficient (standard error)
Level of respondent	
Age category	-0.29 (0.10)
Respondent uses cocaine mainly in work circuit	1.59 (0.51)
Respondent uses cocaine mainly in hard drug scene	1.51 (0.42)
Level of relationships	
Nominees in hobby/sports circuit	-0.91 (0.38)
Duration of cocaine use of nominee	0.13 (0.05)
Respondent in interaction with relationships	
Tertiary education level respondent, nominees in work circuit	-0.81 (0.38)
Gender of respondent male, gender of nominee female	-0.78 (0.24)

Table 7.10 shows that, on the level of the respondent, the age of the respondents has a weakening effect on the role of cocaine in the relationships. This means that the relationships of older respondents are less in the context of cocaine than those of the younger respondents. Cocaine appears to play an

important role in the relationships of respondents who have stated that they take cocaine mainly at work. This refers to four respondents who all work in pubs/cafes. The average scale value for relationships of these respondents is 3.7 (the overall average for all relationships is 2.2). Cocaine is also very important in the relationships of respondents who take cocaine mainly in the hard drug scene. The average scale value for the relationships of the respondents in the hard drug scene is 3.9.

On the level of relationships, it appears that cocaine plays a minor role, as far as the procurement of cocaine is concerned, in the relationships of respondents and nominees in the hobby/sports circuit. The average scale value for these relationships is 2.1. The duration of cocaine use of the nominees has a reinforcing effect on the role of cocaine in the relationships. This means that to the extent that a nominee is using cocaine for a longer period, the role of cocaine in the relationships is more important than in the relationships in which the nominee has been using cocaine for a shorter period.

Furthermore, significant effects were found between characteristics of the respondents in interaction with characteristics of the relationships. When a respondent has followed tertiary education and he nominates someone in his work circuit, it appears that the role of cocaine in this relationships has little significance. The average scale value for these relationships is 1.6. It seems that in these work situations people know that the other is using cocaine but cocaine is a side issue in the relationship. In relationships in which a male respondent nominates a female user, the role of cocaine appears to be of minor importance.

7.3.2 Chain analysis Barcelona

In Barcelona much effort was made in extending the snowball chains. Analysis of the three main chains gives a broader vision of the relationships within these chains, and, above all, of the interrelationships within the chains of users of different types. The first chain is composed of groups of friends who are highly interrelated, from a neighbourhood of Barcelona, with a consumption level that is in general low. The chain was composed of social types. The second chain, with less connected relationships, consists of personal networks of individuals who are also young, but less so than in the previous chain, with relationships that are more centred in going out at night and a higher level of use. The third consists of individuals who live in, or are connected with, a small town close to Barcelona. In this chain, as central nucleus and more interrelated, appear various individuals with a very high consumption and also with involvement in cocaine trafficking.

These three chains can contribute a characterization of three different lifestyles that assume, moreover, a gradient in the level of consumption and in the importance of cocaine within the lifestyle: the first chain, is called the group of friends, and has a low level of consumption; the second, that of the moderns, has an average level; and the last, the dealers, has a high level of consumption. With these denominations we wish to emphasize, above all, the type of relationship that characterizes the dense nucleus of the chains.

In the first chain, the group of friends, the relationships are very closed. Almost everyone knows everyone else, and the relationships occur preferentially within the neighbourhood. In the second chain, the moderns, more open relationships occur: there are many of them and there is less acquaintanceship with others in the chain. Relationships occur especially at night, in bars, pubs or discotheques in the city. Finally, the third, the dealers, is also a chain of open relationships, but the nucleus of habitual users and those that deal (at whatever level) is very compact. The relationships in this dense nucleus occurs in the locality of the metropolitan area of Barcelona where they live.

In the three chains, there appear individuals who are cited in all or almost all the personal networks; moreover, they are also those who have a higher level of consumption and a more direct relationship with the market. In all cases, there also appear individuals of the social type. The number of these consumers is in the majority in the first chain, and minimal in the third. In all three chains the situational type is also found, while the commercial type is only found in the second and third chain. Especially in the third chain both the users that perform in traffic between friends as well as those who are deeper involved in the market are important individuals in the chain. They are the users with a higher level of use.

In the first chain (the group of friends), most are middle managers and students; in the second (the moderns), the number of those that work in the world of fashion, advertising, etc, and in the night scene, is high; and in the third (more homogeneous), those that work in the night scene and those that are engaged in illegal activities are prominent.

There also exists a gradient in the three chains, if we analyze the circles of use. Both the respondents, as well as the nominees in the first chain, use the drug preferentially in the entertainment circuit. The individuals in the second chain also use the drug mostly in public places, but a quarter also use it in private circles. Lastly, in the third chain, it is consumed above all in private and public circles, and the number of individuals who consume in any place and situation (almost a quarter of the nominees) is relatively high. Therefore, the preferential consumption in public places of the friends of the first chain, is at the other extreme of the more private type of consumption of the dealers in the third chain.

The data that refer to the sort of relationship also seem to reflect some characteristics that differentiate the three chains. The major relationship in the first chain is that of friendship or intimate friendship; in the second chain, the relationship of intimate friendship decreases very strongly; and in the third, the relationship of acquaintanceship (almost a quarter) is prominent. The relationship of intimate friendship is equally low. In the first chain, we can see how the major relationship with the nominee is prior to the first use of the drug. This proportion decreases in the second chain, where less than a quarter of the respondents have a relationship before the first consumption. A similar situation appears in the third chain. The frequency of contact in the last six months, in the group of friends (first chain) is higher than that of the moderns. Also quite high is the frequency with which more than half of the individuals of the third chain see each other; but, on the other hand, close to a quarter have had little contact.

The contacts that occur with more frequency in the context of use/consumption are to be found in the second chain (the moderns). The respondents in the chain group of friends, and those of that of the dealers, have a very frequent relationship, but the least part of the contacts occur in contexts of consumption. The relationship of friendship and the festive consumption of the first chain, supports this situation. It is also meaningful in the case of the third chain, although for other reasons: many of the contacts have a commercial purpose (legal or illegal) that is not explicitly related to cocaine and, moreover, there is a high degree of concealment. Only in the second chain (the moderns), exist many users that only have relationships in the contexts of consumption, at night, in bars, pubs, discotheques, etc. Outside of these spheres of use, they do not have any relationship.

Finally, looking at the relationship between respondents and nominees, and the manner of getting the cocaine: in the second chain (the moderns), the majority states that they give or sell cocaine to nominees (asymmetrical relationship). The qualitative information enables confirmation of these data: respondents of the situational type invite, or are invited, with more frequency, and they also deal among friends. In the other two chains, we see a similar situation among them, with an interpretation that is also similar. Most of the constituents of the group of friends and those of the chain of dealers state that they have the same dealer. In the group of friends, those that deal are users of the situational type, who deal among friends (especially one of them, often cited); in the other, they are dealers with a greater level of involvement and consumption (here too, it is easy to locate a respondent of the commercial type).

7.3.3 Total personal network analysis Turin

Turin carried out a second-level analysis centred on the respondents' total personal network ('set', Barnes 1977), focusing special attention on primary ties. This perspective allows us to analyze some structural and interactional characteristics of the networks that would otherwise be incomprehensible. The basic questions to be formulated are: the size of the primary network, or immediate set; its density, the degree of specialization in cocaine use, the connection between the different clusters, and the information about social control by referring to the knowledge of the conditions of an individual as a user.

By means of the ratio between the number of users and the total size of the personal network, we obtained a simple index of the pervasiveness of the substance in the respondent's network. The average personal network of the respondents is constituted of about 14 persons who used or had used cocaine - i.e. 58% of the subjects - and about 10 individuals (42%) who neither used nor had ever used the substance.

With regard to the relational circuits friendship and entertainment the percentage of users is quite high, while it is obviously lower with regards to work and family. On the average the respondents identified as users about two thirds of their friends and almost 75% of the contacts in the entertainment context. The ratio number of users/number of non-users decreases in the relational context of work (49% of the mentioned subjects are users), reaching the smallest percentage in the circuit of familiar relations (10%). Therefore, already from the first interpretation of the characteristics of the networks, emerges a confirmation of the qualitative investigative work: the importance acquired by the friendship and entertainment circuits as preferred environments for cocaine use.

When filling in the graph, each subject was asked to identify, both among the users and non-users, the persons who knew that he/she (ego) was using or had used cocaine. Through this simple question we intended to assess:

   a. the level of reciprocity of the relations among the users;
   b. to what extent the behaviour is known in one's own network, also beyond the part of persons who use; of course, this index must be made contextual;
   c. an indication, though weak, on the depth of the relation between ego and the members of his network, resulting from the knowledge of a practice which is sanctioned in any case, at least on the level of the formal law.

The results seem to underline clearly a few definite points:

   • the relation between ego and the other users in his network is centred on a high level of reciprocity with regards to the knowledge of cocaine use: 70% of the members of the network of users know of the respondents using cocaine, and only the remaining 30% do not know that;
   • although the respondent himself selected them as significant and from the primary relationships, the relations with the other non-users of the network do not seem to be characterized by such an intimate level enabling them to know of ego's using cocaine: among the non-users only 17% seem to be acquainted with ego's use, while 83% ignore it.

Looking at these results we can see that on a general level cocaine use does not appear to be so widespread that use can occur in the presence of persons who do not use it, as was seen in Rotterdam. This strengthens a series of elements perceived in the interview, where, among the rules of the user's behaviour, the rule of non-use with non-users (that is, never with the regulars, as an respondent said) strongly emerged. The low index of knowledge concerning non-users (who build the 42% of the mean personal network) suggests, in the case of Turin, an interpretation of the phenomenon cocaine use as a social but yet hidden phenomenon.

The ratio known/not known in the friendship circuit inverts the percentage in reference to the friends being users or non-users. Such a high percentage (77%) of non-user friends not knowing of ego's using cocaine confirms the hypothesis that use is particularly hidden; i.e. a behaviour to hide also from the great majority of ones own friends who do not share it. Furthermore, no indications of neutrality towards the use of the substance seem to emerge: among non-users ego's behaviour is mostly unknown. Ignorance means not only that cocaine is not talked about but also that the respondent can succeed in controlling his relationship with the substance without being discovered by the other significant relationships, i.e. without displaying such deviant and compulsive behaviours that could betray the respondent's use. This finding is supported by the results of the qualitative section and of the questionnaire.

Further indications on this subject emerge from another structural dimension of the networks: density, i.e. the relation between the ties which are possible on the theoretical level and those actually activated. On the whole, the respondents' personal networks resulted large-mesh, being characterized by a density of about 22%; analogous (about 23%) to the value of the density in the entertainment circuit. In the friendship circuit the average density is much higher (51%); in this case it can be referred to as an averagely large-mesh network. In this circuit the relations among users mostly (over 30%) assume the form either of cliques, that is small groups 100% interconnected, or averagely interconnected clusters (34% of the cases).

About one third of the networks in the friendship circuit is a large-mesh, has a density lower than 30%. This averagely close/very close structure is a corollary of the type of use mostly occurring with other users; the small group (in the friendship circuit) seems to emerge from this high level of density as a priority form. The low degree of density in the level of the general personal network point to a very labile network of interconnections among users in the different relational circuits (just large-mesh). This implies direct contact between ego and the other users, but no direct ties between the users through different circuits. A number of the analyzed networks assume a segmented structure, i.e. made up of a union of several complete sub-graphs that are mutually isolated. Within the subgroups all possible ties are present, but between the subgroups there are no ties (Baerveldt and Snijders 1991). In the whole personal network, use seems spread - in different proportions - across the various relational circuits but cocaine does not appear to be an adhesive strong enough to establish ties between all the users.

7.4   Estimations of extent

The number of cocaine users has been estimated by means of one-wave snowball sampling. The first two methods were developed by the research team in Rotterdam, based on the capture-recapture method (Snijders 1991). The third, developed by the Barcelona team, is based on the likelihood function (Del Castillo 1991). The first two methods, and the third, are complementary. The first two are useful when the number of respondent-referrals, i.e. the number of persons referred to on level zero by other persons on the same level, is not excessively low. The third method is useful when this number is zero or very small.

An attempt was made to use the same snowball sampling procedure in all three cities, but its evolution was different. The percentage of people referred to by the persons on level zero differs considerably from one city to another. The amount of difficulty involved in locating nominees also varied. One explanation for this is the different legislation in force in the three countries. Higher tolerance in the use of cocaine corresponds to being easier to refer to other users and to a higher rate of success in finding them. Another important aspect is the phenomenon itself with all its cultural and socio-economic aspects. For example, in a study of heroin users in Groningen in the Netherlands, it was shown that respondents were more than willing to mention other heroin users and to help in arranging an interview with the selected nominees (Intraval 1991).

Rotterdam

As stated, the number of cocaine users in Rotterdam is estimated with the assistance of two estimators[6]. Both estimators employ data obtained from a snowball sample of the populations. The estimators are based on a random one-wave sample. The randomly selected respondents form the initial sample. The number of initial respondents is denoted with N. The respondents then nominate other people of the population who meet the inclusion criteria. This collection of newly nominated users forms the so-called first snowball wave. In the Rotterdam survey, the respondents who were originally approached form the initial sample. The other cocaine users whom they nominate must meet the following inclusion criterion: they must have used cocaine at least 25 times and/or at least five times in the last six months. Furthermore, the nominees must live in Rotterdam. The people nominated by the respondents who were originally approached, form the first snowball wave.

The first estimator v_1 is defined as:

$$\hat{v}_1 = \frac{(n-1)T_{01}}{T_{00}} + n$$

in which T_{01} is the number of newly nominated users by the initial respondents, thus if a user has been named more than once he is also counted more than once,

T_{00} is the number of times that respondents nominate other respondents, the so-called respondent-referrals.

The second estimator v_2 is defined as:

$$\hat{v}_2 = \frac{(n-1)M}{M_{00}} + n$$

in which M is the number of newly nominated users by the initial respondents; if a user has been named more than once he is counted only once,

M_{00} is the number of respondent-referrals in which respondents nominated more than once are only counted once.

Both estimators can be understood in terms of the capture-recapture method in which capture is interpreted as drawn in the initial sample and recapture as nominated by users from the initial sample, in other words the respondent-referrals[7]. As we have already said, both estimators are based on a random sample. This is an important assumption which cannot be fulfilled in the Rotterdam study. Cocaine users form a hidden population and, moreover, there is no sample frame available. We have therefore tried, by means of various targets, to find initial respondents who are as far removed as possible from each other. The degree to which this variation, in principle intended as an approach to a random sample, has succeeded, depends largely on the degree of independence of the search for the various respondents by the field workers. If, for example, two respondents are found in the same pub on the same evening or if this pub appears to be the habitual pub of both persons, the chance is great that they nominate each other. This chance is smaller when two respondents are found in different pubs, geographically far apart.

In Rotterdam, a total of 110 respondents were interviewed. 84 of these 110 form the initial sample (the other 26 are extensions). Nine of the 84 initial respondents did not cooperate in nominating other users. Three of these have been nominated by other initial respondents, these three are considered nominees. The same applies to five other respondents who did cooperate in nominating other users, but were found to be non-random (e.g. through acquaintances of the field workers). We are therefore left with n=70 respondents and M=824 nominees for the first estimates.

estimator	estimate	standard error
v_1	2,278	(505)
v_2	2,777	(629)

These estimators are low compared to the estimates from other sources (Intraval 1989, Toet and Geurs 1992). Closer analysis shows that this is due to the high number of respondent-referrals among the respondents. This is an indication that, despite the precautions taken, the initial respondents were not all found independently of each other. Diagram 7.3. shows the initial sample after the high number of respondent-referrals has been analyzed in more detail.

Diagram 7.3   Subdivision respondent-referrals over the entries in the sample

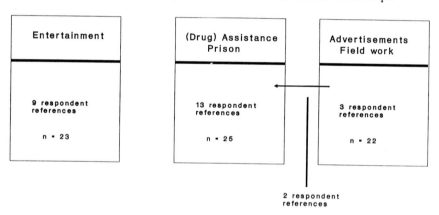

There appear to be two distinct entries. Respondents found in the entertainment circuit name other respondents also found in the entertainment circuit. Respondents found through the assistance agencies or prison together with respondents found through advertisements or through the field work form a single entry because two respondents from the entry advertisement/field work nominated two other respondents from the entry assistance/prison.

In order to arrive at reliable estimates, it is necessary to check which respondents in the initial sample were not found independently of each other. If, for example, a respondent has said to a field worker 'you should take a look there' and a new respondent is indeed found, then this new respondent cannot be considered as having been found independently, but as a selective extension. Another sort of finding a respondent in a non-independent manner is when two respondents come together for an interview or are found together in the field. 24 respondents appear to have been found non-independently of each other. In order to meet the assumption of a random sample to a reasonable degree, it is necessary to eliminate one of the simultaneously found respondents by drawing lots. Estimators can be calculated for the limited dataset thereby obtained. In this way they were ten times estimated with ten populations in which the respondents who were not independent of each other were excluded. If two respondents A and B were found in a non-random manner, then in the first estimate population, for example, respondent A is excluded, in the second estimate population respondent B and in the third respondent B, and so on. This happens with all groups in which the respondents are not found independently of each other. The averages for the estimators are:

estimator	estimate	standard error
v_1'	8,675	(4,754)
v_2'	8,545	(4,620)

Another way to estimate the number of users is the estimate per entry. From diagram 7.3 it appears that there are two separate entries. With the 23 respondents from the entertainment circuit, it appears that ten respondents were not found randomly/independent of each other. Ten estimates were made for this entry in which the respondents who were not found independent of each other were excluded alternately. The same was done for the respondents who are found by means of the other entry (assistance agencies and prison, with advertisements and field work). It turned out that 14 of the 47 respondents selected in this way had not been found independently of each other. The averages for the estimators when the two entries are added to each other are:

estimator	estimate	standard error
v_1''	8,833	(3,596)
v_2''	8,701	(3,320)

It appears that both approaches, i.e. for all randomly selected initial respondents and for the two separate entries arrive at estimates of around 8,700 users. The high standard errors of the estimators indicate the degree of inaccuracy of the estimates. We have several reasons to believe that these estimates are in fact underestimates. The first reason is that the initial respondents were found too much in the centre of the network. This is because respondents in the centre of a network have a higher chance of being found since they know more people and are known to more people. This would result in a large number of respondent-referrals. The second reason is that the inclusion criterion contains, among the other things, have used cocaine 25 times or more. This means that users who meet the criterion but took cocaine in the distant past and stopped some time ago, will probably not be found and will not be nominated.

In addition there is a standard rule in statistics that an estimated parameter normally lies within two standard errors of the estimator. Because we have reasons to believe that the above estimates are underestimates, we could suppose that there are at least 5,000 cocaine users in Rotterdam who meet the inclusion criteria. The maximum number of users would, according to the rule, fall within two standard errors of the estimates. But since we do not know the degree of underestimation it is possible that the maximum number of users is 20,000. If we accept that the random sample is of a high quality, with sufficient attention being paid to the spread of the initial respondents and that the ex-users from the distant past form only a small group, we may conclude that the best possible estimation of the number of cocaine users who meet the inclusion criterion would be around 12,000[8]. This is, incidentally, 2% of the overall population of Rotterdam (and 3% of the age category 15-50 years).

Barcelona

According to the data obtained in the socio-epidemiological survey on the use of drugs carried out by the Generalitat (Autonomous Government of Catalonia) in 1990 (see Departament de Sanitat 1991), in the region of Catalonia 3.3 per cent of the population surveyed in the 15-64 age group stated having tried cocaine at some time (lifetime prevalence). This percentage falls to 1.7 per cent stating having tried it in the last year (last year prevalence). If the ratio is limited to the city of Barcelona, applying those rates to the data from the 1991 Census, we find 37,000 persons who have taken cocaine at some time (including experimental users); of these, 24,000 used it in the last year. We only use these data as a reference to estimate the extent (magnitude) of cocaine use in Barcelona.

With regard to the data that we provide here, two preliminary considerations have to be made on the characteristics of the consumers on whom the estimation of the user size has been calculated. In the first place, the criterion of inclusion used is more restrictive than that of the survey quoted above. In our case, the cocaine users interviewed, and those referred to by them, have an established pattern of use (even though this means minimum use, experimental users remain excluded from the sample). In second place, ex-consumers are included in the total sample of users and also of persons referred to. Among the respondents, ex-users represent seventeen per cent and among the persons referred to, eight per cent. In both samples, for eight out of ten of them, the most recent period of abstinence had a duration of a maximum of one year.

For this reason, for the purposes of estimation, their status of ex-users has not been considered.

On the basis of an initial sample of fifty cocaine users, a snowball sampling procedure has been carried out, in such a way that each of them has referred to a certain number of acquaintances who also use cocaine. The number of persons referred to for each of these fluctuates between zero and twelve. Thus a total of 225 persons referred to has been acquired. Despite the fact that the snowball sampling continued at more levels, for the purposes of this section, it will not be taken into account. After carrying out the above sampling, we observed that no individual is repeating. That is to say, that the 225 persons referred to and our initial sample of fifty users give a total of 275 different users. This indicates to us, intuitively, that the size of the user population will be much greater in comparison with these 275 that we have differentiated.

The way that we have developed our data provides us with an estimation of the minimal size of the user population. Intuitively the idea consists of looking for a size of the population N_0 such that, if the real population size is above this value, then the probability of not finding any repetition in a sample such as we have is very low (for example, < 0.05). We can calculate the probability that no repetition is produced as a function of the size of the user population N (likelihood function). It is important to note that we then only find the minimum number of the research population. An assumption is that each individual selects the persons referred to as if he were making choices at random within the user population. However, although we believe that this does not correspond to reality at all, it is a frequently used working hypothesis. Below, the graph shows us the probability function in which we have marked the values corresponding to $p = 0.05$ and $p = 0.1$ (figure 7.1).

In fact what we are doing is looking for the likelihood intervals of five per cent and ten per cent respectively. This way then gives us only an estimation of the minimum number of cocaine users in Barcelona of approximately 12,000, with a confidence level of 95% This means of course, that the total number of users will be higher. Bearing in mind the volume of the population in the city, this figure represents a minimum of approximately 1% of the population over the age of 14, and of 2% of the population between 15 and 44 years of age. If we choose a confidence level of 90%, this minimum value increases to 15,500 users[9].

Figure 7.1   Probability of no repetition being produced in function with the population

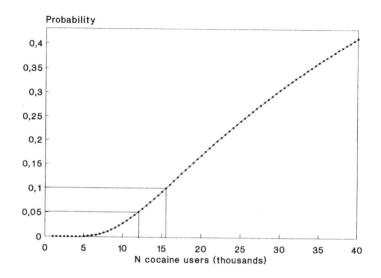

Turin

Unfortunately, in Turin none of the conditions necessary for using the estimators which had been developed could properly be met. The clandestine condition to which the cocaine user is forced in Italy makes it difficult to obtain the necessary information for identifying the nominees in the personal networks, so that in many cases it is not possible to know the size and characteristics of the personal network. Therefore, it is not always possible to distinguish the nominees from one another and to calculate the number of repetitions, that is the number of persons mentioned more than once by different respondents. For these reasons it was not possible to estimate the size of the population of users in Turin by either the Rotterdam or Barcelona estimates.

NOTES

1. In chapter five it has been noted that only a small number of users buy their cocaine in entertainment circuit. The difference with the results here can be explained by the way cocaine is often bought. One person buys for several friends.

2. The circuit here is defined differently than in chapter five. In this section we refer to the most important circuit of use during the whole cocaine career.
3. The scale is constructed on the basis of the following questions:
   a) Does the contact with the nominee occur in the context of cocaine?
   1. Always; 2. Usually; 3. Occasionally; 4. Never.
   b) Does the nominee know that you take cocaine?
   1. Yes; 2. No; 3. Not known.
   c) Have you and the nominee been linked in any way in obtaining cocaine in the last 6 months?
   1. Yes, I obtain/sell from nominee; 2. Yes, nominee obtains/sell from me; 3. Yes, we obtain and sell in turn; 4. Yes, we obtain/sell from a third party; 5. No, there is no connection between us, but we have the same dealer; 6. No, there is no connection of any kind between us.
4. The 95 respondents nominated a total of 439 contacts. At the moment of the interview, 32 respondents were not taking cocaine. These respondents were not included in the analyses since there is no evidence of a cocaine relationship. The other 63 respondents nominated a total of 309 contacts. In 14 of these relationships it appeared that the nominee does not know that the respondent is taking cocaine. These relationships, too, were excluded from the analyses. The relationships which were investigated are all those in which the nominee is aware that the respondent is taking cocaine. They are, thus, mutual relationships as far as the knowledge of each other's cocaine use is concerned. A total of 259 relationships of 63 respondents remain.
5. The different relationships which have the same respondent are not independent of each other. Therefore it is not possible to carry out a normal regression analysis. Such an analysis is based on the independence between the relationships of a respondent. The relationship level is nested in the respondent level. For an introduction into multi-level analysis see, for example: Bryk, A.S. and S. Raudenbush (1992): Hierarchical linear models for social and behavioral research: applications and data analysis ways. Sage Publications, Newbury Park.
6. For a more detailed description of these two estimators, see the article to be published: 'Estimation of hidden populations by using snowball sampling', by O. Frank (University of Stockholm) and T.A.B. Snijders (University of Groningen). A number of estimators are discussed in this article.
7. The first and second estimator are closely linked. When the average number of times, that a person in the initial random sample is nominated by others in the initial random sample, is equal to the average number of times that an individual outside the initial random sample is nominated by people in the initial sample, then T_{00}/T_{01} is equal to M_{00}/M and both estimators have the same result. The identification of nominees is very important, but often not easy to carry out. Respondents have difficulty naming other users for a number of reasons. Therefore, in the survey we requested the first two

letters of the first name and surname, or nickname, age category and occupation. The people in the first wave were classified on the basis of the combination of these characteristics. This is no easy task as can be seen from the fact that with the 1,041 new names nominated by the 85 initial respondents there are the following percentages of missing data: letters first name 2%; letters surname 32%; gender 0.4%; age 0.4%; occupation 9%. It turned out that in nearly one third of the cases the first two letters of the surname were missing and in nearly one tenth the occupation. These two, together with the initials of the first name, are the best distinguishing factors. For this reason v_1 is more reliable than v_2 since in this estimator only the respondents need to be identified. More data are available on the respondents.

8. It must be noted that the limits (5,000 to 20,000) indicate what can be concluded about the number of cocaine users only on the basis of the statistical analysis of this snowball sample. Any other reliable evidence which may become available can be used to adjust this estimation and make it more accurate.

9. Introducing a slight modification of this method we can also calculate the maximum likelihood estimator in a case where some repetition is produced between the subjects referred to by the components of the initial sample. Applying this to our sample, and supposing that there had been a single repetition, this estimator of the total number of users would give us, with a confidence level of 95%, a maximum value of 35,000. That is to say, according to this assumption, in Barcelona there could be a maximum of 35,000 persons who have taken cocaine at least five times in the last six months and/or 25 times in their lifetime, this representing a maximum of around 3 % of the population over 14 year of age, and 5% of the population aged between 15 and 44.

PART III

DISCUSSION

B. Bieleman, A. Díaz, G. Merlo, Ch. D. Kaplan

CHAPTER 8

SUMMARY AND CONCLUSIONS

The study presented here is a unique and coordinated effort of a multi-disciplinary team of scientists and city politicians to increase knowledge of the real extent and nature of cocaine in Europe. It has often been said that in the case of new drug epidemiological trends, Europe follows the United States, with a five to ten year lag. Is this commonsense knowledge or is it a dangerous myth used to incite panic and promote political interests and ambitions?

This book describes the results, conclusions and recommendations of an intensive study of cocaine use in the cities of Barcelona, Rotterdam and Turin. The aim of the study was to provide a rational, politically relevant, contribution to the assessment and reduction of the harm linked to the drug that appears to have captured the imagination of Europe. Cocaine is a drug with a multiple personality. It has many faces and it is rooted in all strata in society. The intention of this book is to present a soundly based evaluation of the true nature and extent of cocaine use in Europe. The method to achieve this evaluation was the study of a broad-based sample of cocaine users in three, culturally diverse, European cities.

A new way of conducting policy-oriented drug research is presented and a novel methodology explained. This approach proved useful as a means of overcoming the barriers between scientists and politicians which have plagued earlier research in the field of drugs. Multidisciplinary teams were formed, made up of local decision makers, policy planners and independent scientists. Regular meetings were held in each city to discuss interim results and any problems which had occurred. Later meetings were devoted to drafting recommendations and conclusions. This way of working assured the scientists that their work was politically relevant and would not be 'hidden at the back of a drawer', while politicians could see that the financial investment in research would yield a tangible pay-off. They would be provided with a lot of information about the phenomenon. The pay-off for the researchers and politicians is that on the one hand the results of the study have influenced current political and scientific attitudes, and, on the other, they have revealed publicly the facts on cocaine use in Europe.

In all, 363 cocaine users in the three cities cooperated with in-depth interviews. The use of an innovative methodology involving the synthesis of snowball and targeted sampling and network analysis meant that data was collected on an additional 1,635 cocaine users who were the contacts of the interviewed users. It represents the largest specific database, to date, of local community-based samples of cocaine users. This is a remarkable achievement, considering that the entire project was completed in two and a half years, and involved the coordination of three research sites, each in a different country with its own language and distinct social, political and cultural setting.

The main conclusions of this book have an impact on a broad range of issues. The subject, the nature and extent of cocaine use in three distinct cities, poses a complexity of practical and scientific problems which are intrinsic to the study of hidden populations. This would be difficult even on a single research site. Snowball sampling provided the common point of departure for the three multidisciplinary teams in each city. Many of the conclusions reached in the study are original. The one thing we are all aiming for is a better under-standing of the worldwide issue of cocaine use. If this book stimulates debate, discussion and research into the nature and extent of cocaine use in other local communities, it will have achieved its objective.

8.1   Methodology

This book, on the nature and extent of cocaine use in three European cities, deals with the particular complexity of a hidden and rare phenomenon. The research project required a novel approach. The method of snowball sampling was used as the basic premise for a research process centred on diversified methodology. The researchers discarded the traditional opposing claims of quantitative versus qualitative methodologies in favour of a multi-method approach which integrated tools originating in different traditions of research. This combination of quantitative and qualitative methods demonstrated a flexibility which characterized the whole research process.

Operationally, all three cities shared a similar research design centred on fieldwork, snowball and targeted sampling, network analysis and typology construction. The basic knowledge of the circuits and social environments of the target population was obtained through fieldwork. In turn, this formed the basis for contacting the respondents for the interview. The fieldwork establish-ed a special relationship of trust between researchers and respondents without which it would not have been possible to conduct the in-depth interviews. The combination of snowball and targeted sampling and network analysis also made it possible to collect relational data on the contacts of the respondents.

By means of these network data more insight was obtained into the complex relationships between cocaine users, their contacts, and the various circuits in which cocaine is used. Intensive fieldwork, snowball and targeted sampling proved an effective methodology for contacting strata of hidden populations.

The analysis of the research data differed according to the specific context encountered in each city. The combination of personal network analysis and qualitative analysis facilitated an interpretation of the patterns and significance of cocaine use which could not have been achieved by looking only at individual characteristics. The analysis of the network relations between users and the significance assumed by cocaine in these ties is distinctly new. Levels that are often kept separate in theory and empirical analysis, e.g. the levels of macrosocial phenomena and the micro-level of social relations and individual strategies, have been linked in the multilevel methodology of this study. The study was organized on a multi-centric basis with three independent research sites and designed to permit controlled comparisons of results.

8.2   General characteristics

In comparing the characteristics of cocaine users in the three cities, the researchers had to take into account the fact that the samples were not random. However, it was possible to add the population of nominees to the population of respondents, increasing the total of N three to ten times. This enlarged sample made for greater representativeness. Where there were no data available concerning the nominees and the samples could not be enlarged, the distribution of most of the variables was acceptable. On the basis of these procedures, it was possible to obtain a meaningful comparison of the characteristics of cocaine users in the three cities. Overall, the comparative analysis showed more similarities than differences in the general characteristics of cocaine users.

Most cocaine users are men. Cocaine users are predominantly in age range 21-35 years. Their educational background ranges from secondary school to higher education and all types of occupation are represented. The great majority of users started taking the drug before the age of 25, after experience with alcohol and cannabis. Rather than replacing other drugs, cocaine is often combined with them, particularly alcohol, cannabis, amphetamines, XTC and LSD. Part of the respondents also used (or had used) heroin. The level of cocaine use appeared to vary, but there is usually a soft (low frequency, low amounts) initiation that increased considerably in the heaviest period. The patterns of use tend to present one or more peaks but discontinuous, constantly increasing and continuous, same level patterns were also found. People who used heroin in addition to cocaine, took cocaine more frequently and at higher levels than those had who never used heroin.

The most prevalent way of use is intranasal, although during the period of heaviest use the percentage of respondents using cocaine intravenously or by way of free basing increases. Most of the respondents who injected cocaine also used heroin. Free basing is a way of use employed by both heroin and non-heroin users. In Rotterdam cocaine is administered by way of sniffing, injecting, basing and chasing the dragon whereas the Barcelona respondents sniff, inject and base cocaine. In Turin, cocaine is only taken by way of sniffing and injecting.

The most prevalent circuits of use during the initial period are the intimate and entertainment circuits. Each city showed differences relating to other periods. In Barcelona, there was a spread into other circuits while in Rotterdam the hard drug circuit became more important. In Turin the entertainment circuit emerged as more important after the initial period.

Cocaine use and criminality do not necessarily go hand in hand. Only a few people engage in criminal activities linked to cocaine use though there is an increased tendency in the period of heaviest use. Involvement with dealing in the cocaine market is limited. The perception of users of cocaine-related problems varies greatly and few make use of services to assist drug addicts. In Rotterdam more respondents have contact with these services than in Barcelona and Turin. Cocaine users who also use heroin have more serious problems than those who have never used heroin.

8.3 Typologies

A deeper examination of the nature of cocaine use required a typology construction. This method goes further than the usual description of cocaine users found in the literature. The common research design employed in the three cities allowed for the development of different approaches to the complex work of typology construction. As the analysis proceeded, the significance of cocaine within the lifestyles of people became an important theme. Cocaine has various meanings and can play many different roles. It can range from peripheral to central in a user's lifestyle. Heroin, on the other hand, has a greater overall impact on lifestyles.

The socio-economic status of users does not play any significant role in cocaine use and cannot be employed as a means of predicting problems. With cocaine it seems that it is not so much the substance itself but the significance attached to the drug that is relevant to the occurrence of problems. The significance of cocaine within each lifestyle is linked to the way it is used and the level of use. These, in turn, are strongly related to problematic aspects. Taking all of this into account, great variations in cocaine lifestyles can be distinguished. These variations can be grouped into two large categories in which the perception and use of heroin plays a major role.

One category of cocaine users also use heroin (Poly-drug Type). They are part of the heroin subculture where the use of drugs is a central feature of the lifestyle. The consumption of cocaine generates serious problems both for the individual and the community. Problems increase and become more severe when cocaine, used intravenously or by way of basing, replaces heroin.

Another category of cocaine users (the majority) do not use heroin. Within this category, there is great variation regarding significance of use and problems with use. First we can distinguish the Leisure Type, a group of users who are well integrated in society and use the drug only in context of leisure activities. In addition to cocaine, they also use legal drugs (alcohol, tobacco) and/or illegal ones (cannabis). Problems directly linked to the use of cocaine are rare or non-existent. Secondly, there is the Instrumental Type, a group which, in addition to consuming cocaine in leisure situations, also uses cocaine as an energy booster at work or for other purposes. This type of use can generate definite problems (physical, psychological or economic) for some users. Most are capable of solving these problems without outside help. Finally, we can distinguish the Cocainist Type, a small minority of users. This group has serious problems. Cocaine has become the central element in their life.

A clear distinction must be made between cocaine users who also take heroin and those who use only cocaine. These two groups inhabit totally separate worlds. Heroin and, above all, the rejection of the needle as dangerous are the elements determining the exclusion/inclusion in each category. The needle is seen as a symbol of social rejection, delinquency, AIDS, and death. This perception is often shared by those who are addicted to heroin and who see themselves as rejected by non-addicts. The way of use draws a clear line of distinction between the categories: intranasal (non-heroin) and intravenous (heroin). In this sense it very important to emphasise that basing as a way of use, is different. It does not have the same negative connotation as injecting and is not linked to the so-called junkies. Most users in the non-heroin category who have serious problems with cocaine, use basing as the main route of ingestion. Basing may be considered as dangerous as injecting because in a relative short time it can lead to compulsive consumption and related problems. Due to their lack of a negative attitude towards basing, the Instrumental Type, and those grouped under the Cocainist Type who are sniffing, should be considered as a risk group.

To summarize, the intranasal use of cocaine is the most common way of use and only a minority have serious problems related to use (a small proportion of the Cocainist Type). The compulsive and problematic use of cocaine is more related to the intravenous consumers of heroin and free basing. Non-heroin users who have basing as the main route of ingestion run the risk for cocaine-related problems.

From an intervention point of view it is important to take account of two distinct groups. The group of poly-drug users, most of whom have problems with heroin and with cocaine can usually be reached through already existing drug assistance agencies. The second group, who do not use heroin and of whom the large majority have no problems with cocaine, usually solve their own problems and do not need any form of assistance. Only a few have serious problems and need outside help. This group will avoid services provided for the poly-drug group since they do not wish to be identified as junkies. They can be reached only by drug assistance which is specifically designed for them. Prevention efforts should be directed towards risk groups such as the Instrumental Type. For this group, cocaine has an instrumental meaning which is related to more habitual use and risky route of ingestion (basing).

8.4   Spread, dispersion and extent

Cocaine use is widely spread in the three cities and occurs among different age-groups, work categories and in various locations. In all three cities, cocaine is closely related to the social dimensions. Having a definite symbolic value, cocaine often enhances socialization and is used mainly in social settings and convivial environments. The close connection between cocaine and the leisure/entertainment and nightlife world is seen all three cities. Certain differences were noted, however. In Barcelona and Rotterdam, cocaine is used in public locations (though in specific areas which assure some privacy of use), while in Turin use is mainly in a private environment with close friends. In all three cities, preference for private use before going out for an evening is linked to the wish to avoid dangerous situations. Some respondents used cocaine to enhance their work performance but consumption in the workplace is not high in any of the three cities.

Most respondents had easy access to cocaine. The practice of offering cocaine to other users is widespread. Women, in particular, are more likely to be invited to use cocaine whereas they rarely provide it for others. The procurement of cocaine without going outside the circuit of friendly contacts (traffic between friends) was frequently seen in all three cities. The relationship between user and dealer can also be characterized by mutual trust, discretion and the absence of conflict.

The analysis of personal contact networks carried out in the three cities reinforces the findings of the qualitative analysis. The spread of cocaine use has a relatively open nature. The respondents said that their nominees (the people they named as users) use cocaine mainly, but not exclusively, in the leisure/entertainment and private circuits. Some nominees used it in the work circuit. Social relations among users are not, in the main, characterized by the use of cocaine. Although there is a high degree of mutual awareness of cocaine use between respondents and nominees, cocaine is not a social adhesive which in itself establishes stable social ties between users. Social contact between respondents and nominees is rarely characterized by the presence of cocaine. Thus, the relationships between cocaine users seen in the study (the social networks) are built upon primary ties specified by such variables as age, friendship, circuit and heroin use. Even the Cocainist Type, for whom cocaine plays a dominant role in his lifestyle, continues to function in other relationships and often uses cocaine socially, as well as alone.

It is difficult for researchers to estimate prevalence in hidden populations. In social research, the primary problem is the definition of a case, i.e. the unit on which to base the estimate. This unit often presents difficult and uncertain specification problems. In order to overcome this problem in our study, strict inclusion criteria of use five times in the last half year or 25 times in a lifetime were used. However, an important methodological achievement has been to develop new estimates derived from a combination of snowball random samples and the network approach based on the well documented capture-recapture or likelihood function methods. In Rotterdam, the prevalence estimates provided the best conditions for the application of the theoretical method of capture/recapture. These two estimators indicated that 12,000 cocaine users were active in the city of Rotterdam (minimum estimated 5,000; maximum 20,000), a 2% prevalence of the general population. In Barcelona, the Rotterdam estimates could not be applied because of different methodological conditions. Barcelona developed its own estimator based on a likelihood function to specify the probability that, given the initial sample of respondents, no repetitions among the nominees would occur. Barcelona estimated that the minimum number of cocaine users in the city lies between 12,000 and 15,000, a minimum prevalence of 0.7%-0.9% in the general population and therefore a higher total prevalence. In Turin, it was not possible to meet the conditions for the application of the method of prevalence estimation used in the two other cities.

8.5   Contextual differences

The socio-legal framework in each country ranges from prohibition of all the psychotropic substances in Italy, to a situation of toleration and de facto legalization of some substances in the Netherlands. Spain occupies an intermediate position. These differences influenced the way in which the study was carried out in the three cities. In Turin, the Italian socio-legal context made it, in the first place, difficult to recruit respondents and, secondly, to obtain information on other users from those who were recruited.

Knowledge about cocaine differed in each country, too. The Netherlands and Spain had already accumulated a large amount of data by European standards, while Italy was at a relative disadvantage. The advantage enjoyed by the researchers in Rotterdam and Barcelona enabled them to orient themselves more precisely and to formulate more specific working hypotheses. The Turin group compensated for a late start by benefitting from the experience of the other two cities. In closing, we note that the sociocultural characteristics of the three societies influenced, on the one hand, the spread and patterns of cocaine use and, on the other, the possibilities of making contact and developing a working relationship between researchers and users.

8.6 Policy

Certain general indications relevant to policy emerged from the research project. In recent years cocaine has been present in almost all environments, sectors and levels of society. It can, therefore, be regarded as an interclass drug. The form of the relationship between user and substance varies considerably and changes over time. Most patterns of use are integrated (or can be integrated) in society. Some patterns of use are connected with friendship or work circuits. They present mainly the intranasal route of ingestion, and do not seem to be either destructive or disruptive. Cocaine use can occur over many years, often in an discontinuous pattern. Many users think that they can interrupt use comparatively easily. Most users maintain that cocaine has not caused them problems of a physical, psychic, social or relational nature. Those who do perceive the risks and problems connected with use are mostly high-level users who are often injecting and who are the least integrated in society.

As far as individual and social risk are concerned, the main determinant appears to be the prior or concurrent use of heroin. The criminality connected with cocaine use is a limited phenomenon. It is difficult to argue that cocaine will replace heroin, for its spread involves new and different strata of the population. Cocaine is, therefore, a deceitful substance. It causes neither dependence nor evident problems as long as it is not abused. It is consistent with the emergent social values: pleasure, efficiency, prestige, success, power. In short, it does not seem that any explosive spread of cocaine abuse as has occurred in the case of heroin is to be expected with concurring deaths from overdose, psychological, physical and social impairment and correlated microcriminality. Europe is not about to experience an epidemic with these disturbing features. The situation would be certainly different if crack cocaine should spread. Crack was not found in the course of the research and seems to be virtually nonexistent in Europe (except Great Britain). The properties of this substance and experience in the USA show that crack tends to spread in the particularly disadvantaged strata of society magnifying pre-existing violent and destructive potentials. Thus, from a policy point of view the prevention of a cocaine epidemic should have as its priority the mobilization of social workers to provide targeted information and to intensify efforts to keep existing social support networks of cocaine users free from new and violent marketing organizations and their specialized products such as crack cocaine.

CHAPTER 9

RECOMMENDATIONS

At the close of this book, it is appropriate that the practical implications of the research and way of working of the study be highlighted in a set of policy and science recommendations. The work underlying the book was never meant to be an isolated academic exercise. Rather, so many people from so many disciplines put in so much effort in order to achieve one simple thing: to produce a document that would make a difference in the way we all think about and act upon cocaine. The recommendations have been written in a positive and optimistic spirit. Indeed cocaine is now widespread in Europe and indeed it should be a matter of concern. Yet as a relatively new and fashionable drug, it requires a set of policy and science responses that are themselves new and consistent with contemporary post modern lives, much, like cocaine itself, are at times intensely exciting and at times horribly depressing. A community response to the diffusion of cocaine in Europe will hopefully be well served by a collective reflection and scenario of action inspired by these recommendations.

9.1 Policy recommendations

Due to the complexity, dynamics and international character of the cocaine problem, a political and technical debate should be promoted in order to design a coordinated international prevention and intervention strategy. This debate should also be promoted at a regional and, above all, a local level in order to adopt consensual lines and to develop their application. The debate should centre on a public health policy based on the assumption that cocaine is already widespread in Europe and has largely been absorbed, with few problems occurring, into the existing European sociocultural context. The problems derived from the use of cocaine need to be confronted holistically in a wider framework of improving the quality of life and health. This approach follows the mandates proposed and set into development in recent years by the World Health Organization.

An aspect deserving deeper discussion is the difficulties and incompatibilities that can arise between police and penal interventions, on the one hand, and, on the other hand, interventions from the public mental health side. The viability of this debate will, to a considerable degree, depend on how the arguments are presented. For example, when the lower levels of the distribution networks are placed under pressure, the resulting market irregularities effect regular cocaine consumers. The possible increase in consumption risks resulting from such pressure needs to be evaluated. In this sense, it is also very important to investigate the consequences of criminalizing the consumers. Another relevant question to be debated is that the distribution of cocaine, despite its illegality, is ruled by market laws. A misplaced pressure against cocaine users may well influence, in an unforeseeable manner, the entire social geography of cocaine and result in an undesirable restructuring of supply and demand.

Subpopulation targeted prevention and intervention

Cocaine use is observed in every social class and strata. It is a drug with different facets and is far less sharply defined than heroin, for example. Researchers have known for some time that cocaine is no longer the privilege of a small elite group in the higher socio-economic echelons. Present day users include opiate addicts and this is part of the reason why cocaine has lost some of its glamour and is no longer considered the champagne or caviar of drugs. Any cocaine policy will, inevitably, be less straightforward than a heroin policy, since it will need to reflect the multifold character of the user population. This means that prevention efforts need to be targeted: each social strata needs a different prevention message. These special features make it advisable for intervention and prevention to be oriented towards the lifestyles rather than the drug itself. At the same time, such intervention and prevention needs to be of different forms and directed specifically towards the individual situations of the different risk groups. Information which is distributed concerning cocaine and its risks should be presented clearly, openly and truthfully. It is just as important to avoid being alarmist as it is to differentiate cocaine's effects and remove the myths surrounding the drug.

The political strategies that underlie these targeted interventions should be based on the typologies and classifications discovered in each city. Different policies need to be aimed at the different needs and capacities of each type. While it is true that cocaine is uniformly illegal in each country in question, use and small-scale dealing in the drug should not be further criminalized. The lifestyle types in which cocaine is peripheral should be strongly reminded that their cocaine use must stay peripheral, and no more than this. For other lifestyle types where cocaine has become more central, a strong effort should be made through information and selective law enforcement to weaken the place cocaine has in their life. Treatment and substitution experimental programmes need to be developed for users whose life has become centred around cocaine.

In short, the use of typologies should be seen as providing useful models for policy-makers enabling them to visualize the people they refer to in their policies. These models must function as necessary antidotes to the sensationalist journalistic reports which stereotype the cocaine user. The models can help provide a realistic picture of groups that need specific help, distinguishing them from those that can be left to their own devices.

The two faces of cocaine

No other drug shows a greater diversity in the ways of self-administration. Cocaine is sniffed, smoked, injected, smeared on gums or genitals, or swallowed. Sniffing is the most widespread method. Many people use cocaine in more than one way. The way in which the drug is taken will largely determine its effect. There is a strong correlation between frequency of use and the amount taken. It is possible to identify two broad categories. On the one hand there is sniffing, in which the effects are generally mild and the cocaine is used socially, generally sporadically, in small quantities. On the other hand there is basing and injecting which usually produce sharp short term effects. The user immediately wants more cocaine. In such cases, cocaine is often taken daily and in large quantities. It is particularly in relation to these latter ways that we speak of compulsive cocaine consumption. Basing and injecting are the ways normally employed by opiate users. Free basing is also popular among non-opiate users. Although the effects to some extent resemble those of injecting, free basing has a more positive image among non-opiate users. Common opinion is if you inject you are a junkie, if you base you are not.

It is clear that when referring to cocaine, we must differentiate between two, more or less, distinctive drugs depending on the way of use. Sniffing frequently takes place in a socially integrated setting, without too much personal or social inconvenience. Basing and injecting are practices associated with lifestyles at risk for cocaine problems. The chance of a user becoming addicted by injecting or basing is quite high. Furthermore, sniffing is not wholly without danger. A number of users whose sole way of taking cocaine is sniffing have developed a compulsive pattern of use.

Harm reduction

Harm reduction which is used in the context of policy discussions about heroin where the harm to the individual and society is obvious should also provide the conceptual framework for cocaine policy. While cocaine does not have the visible harmful effects of heroin, this book clearly indicates that cocaine is not such a harmless drug as has been claimed. Addiction depends not only on the duration of use but, more particularly, on the way of self-administration. Compulsive use occurs among opiate users as well as non-opiate users. Cocaine addiction appears to be different from heroin addiction. People often refer to the psychological dependence of cocaine, whereas heroin is said to produce a more physical form of dependence. Studies, including our own, indicate that a number of compulsive users stop of their own accord when cocaine use leads to an excess of problems. In these circumstances aspects such as social-economic status and social bonds are important. Therefore, the reduction of harm for society and for the users of cocaine must be focused on reinforcing those social bonds of the users which do not depend on cocaine. Intervention should imply a strategy directed towards the minimization of harm.

Every effort must be made to promote those aspects of cocaine lifestyles that discourage potentially harmful behaviours such as injection and basing. The promotion of social bonding behaviour that has formerly occurred in the context of cocaine should be encouraged without the use of the drug. Attention must be paid to the public mental health disorders which may emerge along-side or be self-medicated by cocaine. The reduction of harm from cocaine also implies the provision of psychotherapeutic services to complement social bonding interventions. New forms of streetwork need to be developed which will reach out to specific cocaine lifestyles and, if necessary, provide coun-selling and primary care. These forms of streetwork should stimulate the endeavour of targeted cocaine lifestyle types to mobilize their own resources to prevent the harm caused by cocaine.

Counselling services should also take place in the context of existing alcohol services in order to avoid the stigma of seeking help as a cocaine user. The training of professionals should be a priority in this new harm reduction effort. Social workers should lead these efforts, but they should be supported by a new generation of physicians who are competent to prescribe medications and supervise psychiatric treatment. Since social networks are involved in cocaine use, harm reduction should not be restricted to services for individuals. They should stimulate social mobilization, promotion of healthy lifestyles and an active, community-based public mental health policy.

The wide extent of cocaine use is not the same as an epidemic

In Europe the prevalence of cocaine use is sufficiently high to require political concern. On the other hand, there is also no need for panic. While the numbers of cocaine users seem to be gradually approaching American scales, the problems engendered are not enough to alarm the existing drug control systems. Great Britain seems to be closer to the USA, experiencing, for example, incidences of crack cocaine use virtually unknown in other European countries. The situation cannot be defined as an epidemic akin to a contagious disease. Rather, the spread of cocaine throughout all sectors of society provides new prevention possibilities more closely related to existing alcohol policy than to hard drug policy. Cocaine resembles alcohol in that although it is used widely a relatively small minority of users have problems. Every effort should be made not to intensify criminalizing sanctions on the users which would result in the disruption of the existing social bonds which have kept use under firm control. We found that cocaine is mostly consumed at home and in the entertainment nightlife circuits by people who do not take opiates. The widespread use of cocaine should not be cause for alarm, but be seen rather as a potential for mobilizing a broadbased prevention campaign.

Because of the relatively large extent of cocaine, it is predicted that the number of cocaine users with dependence problems will increase both among poly-drug (heroin) users and those taking only cocaine. This means that an increasing number of users will require the services of (drug) assistance agencies. It is advisable to consider new treatment methods with a more therapeutic nature. In the United States, for example, behavioral therapy combined with pharmacotherapies using antidepressants and dopamine agonists are increasingly being used. Another approach already employed by some social workers is the coping model. This may offer opportunities for targeted intervention. It is important to draw a clear distinction between provisions for the more socially integrated cocaine addicts and those for poly-drug users. The former wish to avoid any identification with the latter.

The increase in the projected number of people needing help should not be equated with the epidemic outbreak of cocaine in the United States in the Eighties. This optimistic forecast is supported by the following results from our study: the consumption pattern of most consumers (intranasal use, sporadic frequency pattern and low consumption level); the secondary role that cocaine plays in their lifestyles, the high price of the drug, and its use being integrated in activities related to leisure and with no significance outside this context. Bearing all this in mind, the existing data indicate that the intranasal use of cocaine does not represent a special risk factor either for the user or for society as a whole.

Political coordination

The results and methodology of this research should be aimed at having a political impact wider than the three cities involved. Innovative resolutions which encourage grassroots political development involving networks of European cities and their international sister cities should be supported and coordinated by international organizations such as the European Commission. This book presents a good example of how science and policy can work together at the international level without losing the sensitivity and relevance to local level variations and contexts. A promising initiative stimulating international cocaine policy-development would be achieved if the policy-oriented research model developed in this study were to be used in other networks of cities.

Monitoring cocaine related problems and supply control

While there is some evidence of aggression and vandalism associated with cocaine, the level in Europe is lower than that documented in American studies. When violence does occur, it is difficult to determine if the cause is cocaine, alcohol, amphetamines, or a combination of these substances. Other drugs, particularly alcohol, are often taken together with cocaine. Cocaine neutralizes the effect of alcohol, to some extent, while at the same time creating thirst. Cocaine use by people who do not also take opiates rarely leads to trouble. Opiate users appear to get more aggressive when using cocaine. American studies have shown that the largest cause of cocaine-related death is homicide. This is directly related to the emergence of a new form of supply: e.g. crack and its characteristic dealing patterns. The current situation is that cocaine dealing at the user level causes little nuisance or violence. The risk of this is increased in places where both heroin and cocaine are sold. An international study should be made on the supply side in order to obtain better insight into how changes in supply influence changes in consumer behaviour and lifestyle. Pressure should be exerted on dealers not to market ready made crack cocaine. The most effective pressure would be to immediately arrest dealers who do this and impose hefty sanctions.

In the context of supply control it is necessary to monitor carefully the products appearing on the cocaine market. In Barcelona, in particular, we found a specific commercial cocaine lifestyle linking supply and demand functions. Effective market control is instrumental in preventing increased cocaine mortality, morbidity and violent reactions. Much of the crime and violence provoking aspects of cocaine are not due to the pharmacology of the actual drug but result from market mechanisms (supply and demand) which intervene. Furthermore, the cocaine which the user buys is never pure: it is often adulterated with harmful additives. The important question is what drugs are being used as additives. It is possible that it is the effects of these drugs which are observed rather than the effects of cocaine. The same applies, incidentally, to MDMA (XTC), a drug receiving increasing attention in Great Britain and the Netherlands. A study in which samples of cocaine (and XTC) are randomly collected and analyzed at user level could provide useful information relevant to supply control.

In short, prevention in the area of supply implies a more active and creative approach than mere interception and reduction. Existing political agencies cannot rely solely on the existing police systems to monitor and control the supply-side of cocaine. It would be useful therefore to repeat the research on a periodic basis. Monitoring trends in the nature and extent of cocaine (and other drugs) would complement and parallel the on-going surveys being conducted in the context of the Pompidou Group's multi-city epidemiological study. We recommend a tri-annual repeat of the research.

9.2 Science recommendations

The international study described in this book reflects a new way of working and a new methodology for the policy sciences. This method of working has succeeded in overcoming the barriers between scientists, policy planners and political decision makers which has bedeviled research in the field of drugs. In the study, multidisciplinary teams of city politicians, civil servants and independent scientists were formed which met periodically, in each city, to discuss results and problems and to draft recommendations and conclusions. The research process benefitted immediately from the different competencies of each profession. In this way, the endemic problem in policy research was solved: scientists were assured that their work had political relevance, and would not be forgotten at the back of a drawer, while politicians could see that their investment in research would have a political pay-off and would provide them with useful information. Finally, the methodology did not force an artificial uniformity on the researchers. The model was sufficiently flexible to allow for differences in approach and definition without losing sight of the common line. This model could be used in research into other politically sensitive topics (e.g. AIDS) and in other cities. It would be useful to construct a multi-site set of applications which could be evaluated for their impact on policy development.

Research designs: social networks and qualitative analysis

In the social, epidemiological, and clinical sciences the conditions of laboratory science are rarely met or even aimed at. The usefulness of these sciences is their practical applications which contribute directly to the stock of commonsense knowledge which is the basis of political and clinical decision-making. What these sciences lack in experimental control, they gain in the capacity to develop complex designs to improve the precision and reliability of their results. The use of snowball samples in studies of drug user populations and a combination of qualitative and quantitative research methods has proved effective. It is important, in such a study, that the initial sample be as random, and as independently selected, as possible. This is difficult to achieve fully in hidden populations. However, this aim can be approximated by means of a careful sample strategy as far removed as possible from the prior inclinations of the researcher. Every effort must be made to improve these designs since they are economical, efficient and effective.

Methodological experiments such as those conducted in the large-sample survey research field should be conducted on the link-tracing research designs described in this book. An experiment that compares the results of a snowball sample with that of a household survey in neighbourhoods with a high incidence of drug use in multiple cities could be conducted to evaluate the different sampling methods.

The essential future methodological question is how to create a multi-stage, multi-level sampling design. It is, especially important in the context of timing to know when to move from one stage (or wave) to another. We also need to know precisely when and where in the research process qualitative and quantitative analysis should be linked. An interesting future exercise would be a study design that would construct typologies of drug users from qualitative analysis and, at a later stage, sample targeted sub-populations or general populations in order to ascertain the quantitative distribution of the types in the population.

Promoting qualitative analysis as a research skill

Qualitative analysis in which the use of drugs is placed within the concept of lifestyles provides an excellent opportunity for research into a phenomenon as diffuse as cocaine use. It is, however, difficult to fit qualitative researchers into multidisciplinary teams. Traditionally, they have been professionally trained to work on their own in a strange field. They have become even more defensive and isolated by critical attitudes (i.e. soft science) and rejection by their colleagues. Specialized training is needed to make the skills of the qualitative researcher available to a wider scientific audience. The technique of typology construction seems to be a promising way to transfer the technology of qualitative analysis to other research contexts. Bridges between qualitative and quantitative science need to be built and the basis of future scientific training workshops should be a of typology construction integrated into intensive fieldwork.

Population parameter estimation

The study has demonstrated the feasibility of estimating the size of populations of drug users on the basis of one-wave snowball samples as long as the initial sample is large enough in relation to the estimated populations, and the people in the initial sample are approached as independently as possible. Further methodological/statistical research is needed to get an optimal random initial sample by using different circuits, settings or social milieus. Furthermore, such an initial sample should be seen more as a type of stratified random sample rather than a simple random sample. New mathematical modelling and estimation techniques should be built on the different estimates discovered in this study. The conditions of their application also need to be described in future methodological research.

The personal network approach

In the analysis of network data in this type of research the personal network is the most useful approach and multi-level analysis the most appropriate method. Analyzing relationship patterns, or social networks, of drugs users or other relevant populations has proved a valuable addition to the commonly used research methods. It produces results which are important for both researchers and policy makers. It must be noted, however, that useful answers to relationship questions will be forthcoming only if trust has been established between the respondent and the interviewer. In the course of extensive interviews, a bond must be built up which enables relationship questions to be answered more readily.

A theoretical revaluation of the epidemiology concept

This book shows that a drug can be widely used while, at the same time, the situation can still not be validly defined as an epidemic. Epidemiology has become an almost sacred cow in the drug field. Both scientific and popular concepts of epidemiology tend to construct the drug problem into something fear-laden and alien. While this concept might apply to heroin, it is of little practical use in analyzing cocaine. It would be a worthwhile theoretical exercise to re-examine the analytic roots of the application of epidemiology in the drug field. The early Anglo-American forefathers of the epidemiological concept of drug abuse were far more cautious than their contemporary colleagues. For them, epidemiology was a special case of a more general concept of social diffusion. In certain situations (e.g. heroin spread), epidemiology may apply, but, the words of the American researcher Hunt, one of these early pioneers in drug epidemiology deserve attention as a focus of a new theoretical discussion: 'The superficial analogy between drug use and diseases is easy to make, but no more useful than most casual similarities. The analogy is imperfect: contagious diseases are involuntary, the result of an organic infection; drug use is voluntary, initially the consequence of a conscious act. Nevertheless, drug use is contagious, in the sense that it spreads by interpersonal contact. Therefore, it may be studied, if not 'explained', using models developed to describe diffusion phenomena' (Hunt 1973).

As Hunt recommends, it is the interpersonal contact quality rather the disease-like symptoms that is the pivotal point of any scientific investigation of the nature and extent of the use of drugs in the community. We must restart our thinking with his words clearly in mind. Cocaine is widely diffused in our communities because it resembles more a social innovation than it does a disease. Prevention and intervention cannot forget this simple yet profound insight. Using commonplace, outworn and stigmatising terms even if they are derived from medical science does not help improve the situation. This book will have been a success if it can truly contribute to the reframing of our thinking and action beyond the relative truth of epidemiology and toward a fuller understanding of the way cocaine has become a significant part of our times that cannot be ignored.

LITERATURE

Adler, P.A., P. Adler (1987):
Membership roles in the field research. Sage Publications, London.

Alvira, F., D. Comas (1990):
Consumo de drogas en el muncipio de Madrid. Plan municipal contra las Drogas. Ayuntamiento de Madrid, Madrid.

Anthony, J.C. (1991):
The epidemiology of drug addiction. In: N.S. Miller (ed.): Comprehensive handbook of drug and alcohol addiction. Marcel Dekker Inc., New York, 55-86.

Anthony, J.C. (1992):
Epidemiological research on cocaine use in the USA. In: Cocaine: scientific and social dimensions. Wiley, Chichester (Ciba Foundation Symposium 166), 20-49.

Aranzadi (ed.) (1945):
Código Penal. Aranzadi, Pamplona.

Arif, A. (ed.) (1987):
Adverse health consequences of cocaine abuse. WHO, Geneva.

Arlacchi, G. (1986):
Analisi del mercato delle droghe pesanti. Dipartimento Scienze della Politica-Sociologia, Università di Firenze, Firenze.

Arlacchi, G., R. Lewis (1989):
Il mercato dell'eroina a Verona e provincia. USSL 25, Verona.

Arnao, G. (1983):
Cocaina: Storia, Effetti, Cultura Esperienxe. Feltrinelli, Milano.

Arnao, G. (1991):
I rischi della cocaina. Il Manifesto, 8-5-1991.

Avico, U., C. Kaplan, D. Korczak, K. van Meter (1988):
Cocaine epidemiology in three European communities. A pilot study using a snowball sampling methodology. IVO, Rotterdam.

Baerveldt, C., T.A.B. Snijders (1991):
On the influence of segmentation of networks: theory and testing. (Abstract) Second European Conference on Social Network Analysis, Paris, June 20-22, 1991.

Bagnasco, A. (1986):
Torino, profilo sociologico. Einaudi, Torino.

Barley, N. (1983):
The innocent anthropologist. Notes from a Mud Hut. British Museum Publications Ltd, London.

Barnes, J.A. (1977):
Class and committees in a Norwegian Island Parish. In: S. Leinhardt (ed.): Social Networks. Academic Press, New York, 233-252.

Barrio, G., J. Sánchez, L. de la Fuente. (1990):
Cocaína en España, 1984-1989. Indicadores de oferta y consumo. Comunidad y Drogas, 15, 9-36.

Barrio, G, J. Vicente, M.J. Bravo, L. de la Fuente (1992):
Consumo de cocaina: evaluación de la situación en España en relacón con la situación en Europa y Norteamérica. Plan Nacional Sobre Drogas, Madrid (in press).

Becker, H.S. (1963):
Outsiders: Studies in the sociology of deviance. Free Press, New York.

Benowitz, N.L. (1992):
How toxic is cocaine? In: Cocaine: scientific and social dimensions. Wiley, Chichester (Ciba Foundation Symposium 166), 125-148.

Bie, E. de, S. Miedema (1990):
Respectabiliteit onder druk? Kwalitatief onderzoek naar levensstijlen van jongvolwassen vrouwen met een uitkering. Onderzoekscentrum voor criminologie en jeugdcriminologie, RUG, Groningen.

Biernacki, P., D. Waldorf (1981):
Snowball sampling. Problems and techniques of chain referral sampling. Sociological methods and research, 10, 2, 141-163.

Bone, J. (1992):
How drug barons carved up the world. The Times, 29-9-1992.

Bott, E. (1971):
Family and social network. Travistock, London.

Bourgois, P. (1989):
In search of Horatio Alger: Culture and ideology in the crack economy. Contemporary Drug Problems, 16, 619-649.

Brussel, G. van (1991):
Epidemiology of drug related deaths in Europe: research issues and preventive implications. IREP, Paris.

Bryk, A.S., S, Raudenbush (1992):
Hierarchial linear models for social and behavioral research: applications and data analysis methods. Sage Publications, Newbury Park.

Budd, R.D. (1989):
Cocaine abuse and violent death. Am. J. Drug Alcohol Abuse, 15, 4, 375-382.

Budney, A.J., S.T. Higgins, J.R. Highes, W.K. Bickel (1992):
The scientific/clinical response to the cocaine epidemic: a MEDLINE search
of the literature. Drug and Alcohol Dependence, 30, 143-149.

Burroni, P., A. Consoli, G. Merlo (1989):
Le tossicodipendenze a Torino: risultati preliminari di uno studio sui fattori
di rischio. Giornale di Neuropsichiatria Età Evolutiva, 9, 1989.

Chitwood, D.D. (1985):
Patterns and consequences of cocaine use. In: N.J. Kozel, E.H. Adams
(eds.): Cocaine use in America: epidemiological and clinical perspectives.
US Government Printing Office, Washington DC (NIDA research
Monograph 61), 111-129.

CIS (1988):
Centro de Investigaciones Sociológicas. Encuesta drogas 1988. REIS, 43,
199-328.

CIS (1989):
Centro de Investigaciones Sociológicas. Encuesta drogas 1989. REIS, 47,
345-407.

Cohen, P. (1989):
Cocaine use in Amsterdam in non-deviant subcultures. Inst. Soc. Geogr.,
UvA, Amsterdam.

Coleman, J.S. (1958):
Relational analysis: The study of the social organizations with survey
methods. Human Organizations, 17, 4, 28-36.

Del Castillo, L. (1981):
Aspectos legales. In: F. Freixa, P.A. Soler (eds.): Toxicomanías un enfoque
multidisciplinario. Fontanella, Barcelona, 455-475.

Del Castillo, J. (1991):
Mostratge i estimació de poblacions ocultes. Qüestiió, 15, 1, 55-68.

Departament de Sanitat (1986):
Enquesta socioepidemiològica sobre el consum d'alcohol, tabac i drogues
illegals a Catalunya. Generalitat de Catalunya, Barcelona.

Departament de Sanitat (1991):
Enquesta socioepidemiològica sobre el consum d'alcohol, tabac i drogues
illegals a Catalunya. Generalitat de Catalunya, Barcelona.

Des Jarlais, D.C., E. Friedman (1989):
Prevalence figures cocaine use New York. Presentation on congress Department of Justice, June 1989, Den Haag.

Díaz, A., M. Barruti, C. Doncel (1992):
The lines of success? A study on the nature and extent of cocaine use in
Barcelona. Laboratori de Sociologia (ICESB), Ajuntament de Barcelona,
Barcelona.

Domingo A., J. Marcos (1989):
Proyecto de un indicador de 'clase social' basado en la ocupación. Gaceta Sanitaria, 10, 3, 320-326.

Eddy, P. (1989):
Het kartel van de dood. Intermediair 1-9-1989.

EDIS (1985):
Equipo de Investigación Sociológica. El consumo de drogas en España. Cruz Roja Española, Madrid.

EDIS (1986):
Equipo de Investigación Sociológica. El consumo de drogas en Aragón. Diputación General de Aragón, Zaragoza.

Engelsman, E.L. (1989):
Dutch policy on the management of drug related problems. British journal of addiction, 84, 211-218.

Epen, J.H. van (1988):
De drugs van de wereld, de wereld van de drugs. Samsom, Alphen a/d Rijn/Brussel.

Epstein, A.L. (1978):
Ethos and identity. Aldine, Chicago.

Erickson, B.H. (1978):
Some problems of inference from chain data. In: K.F. Schuessler (ed.): Sociological Methodology. Josey-Bass, London, 276-302.

Erickson, P.G., E.M. Adlaf, G.F. Murray, R.G. Smart (1987):
The steel drug: cocaine in perspective. Lexington Books, Mass/Toronto.

Erickson, P.G., G.F. Murray (1989):
The undeterred cocaine user: Intention to quit and its relationship to perceived legal and health threats. Contemp. Drug Probl., 16, 2, 141-156.

Escohotado, A. (1989):
Historia de las Drogas. Alianza, Madrid.

Farrar H.C., G.L. Kearns (1989):
Cocaine: clinical pharmacology and toxicology. Journal of Paediatrics, 115, 665-675.

Fibiger, H.C., A.G. Phillips, E.E. Brown (1992):
The neurobiology of cocaine-induced reinforcement. In: Cocaine: scientific and social dimensions. Wiley, Chichester (Ciba Foundation Symposium 166), 96-124.

Fieldler, J. (1978):
Field research: A manual for logistics and management of scientific studies in natural settings. Joney-Bass, London.

Frank, O., T.A.B. Snijders (1992):
Estimating hidden populations using snowball sampling. Submitted for publication.

Funes, J., O. Romaní (1985):
Dejar la heroína. Cruz Roja Española, Madrid.

Gawin, F.H., H.D. Kleber (1986):
Abstinence symptomatology and psychiatric diagnosis in cocaine abusers. Arch. Gen. Psychiatry, 43, 107-113.

Gawin, F.L., E.H. Ellinwood Jr. (1988):
Cocaine and other stimulants: actions, abuse and treatment. New England Journal of Medicine, 318, 1173-1182.

Gawin, F.L. (1991):
Cocaine addiction: psychology and neurophysiology. Science, 251, 1580-1586.

Goffman, E. (1961):
Asylums: Essays on the social situation of mental patients and other inmates. Anchor Books, Garden City, New York.

Goldstein, P.J., H.H. Brownstein, P.J. Ryan, P.A. Bellucci (1989):
Crack and homicide in New York City, 1988: A conceptually based event analysis. Contemporary Drug Problems, 16, 651-687.

González, C., J. Funes, S. González, I. Mayol, O. Romaní (1989):
Repensar las drogas. IGIA, Barcelona.

Goodman, L.A. (1961):
Snowball sampling. Annals of Mathematical Statistics, 32, 148-170.

Grapendaal, M. (1989):
De markt van wit en bruin. Inkomstenverwerving van harddruggebruikers en de rol van methadon. De Psycholoog, 24, 7-8, 357-363.

Grapendaal, M., Ed. Leuw, J. Nelen (1991):
De economie van het drugsbestaan. Criminaliteit als expressie van levensstijl en loopbaan. Gouda Quint, Arnhem.

Grund, J.P.C., N.F.P. Adriaans, C.D. Kaplan (1991):
Changing cocaine smoking rituals in the Dutch heroin addict population. British Journal on Addiction, 86, 439-448.

Hartnoll, R., R. Lewis, M. Mitcheson, S. Bryer (1985a):
Estimating the prevalence of opiod dependence. The Lancet, January 26, 1985, 203-205.

Hartnoll, R., E. Daviand, R. Lewis, M. Mitcheson (1985b):
Drug problems: Assessing local needs. Bisbeck College, London.

Helfand, W.H. (1988):
Mariani et le vin de coca. Psychotropes 4, 3, 13-18.

Hughes, P.H., G.K. Jarvis, U. Khant, M.E. Medina-Mora, V. Navaratnam, V. Poshyachinda, K.A. Wadud (1982):
Modelos etnográficos y de secreto entre los toxicómanos. Boletín de Estupefacientes, 34, 1, 1-3.

Hunnik, M. van (1989):

Je wilt nog niet naar huis dus neem je een snuif: jongeren over cocaïnegebruik. Jeugd & Samenleving, 19, 8, 500-512.

Hunt, W.H. (1973):

Heroin epidemics: a quantitative study of current empirical data. The Drug Abuse Council, Washington DC.

Inciardi, J.A. (1986):

The war on drugs. Heroin, cocaine, crime and public policy. Mayfield Publishing Company, Palo Alto, California.

Interpol (1990):

Cocaine use in Europe 1989. Interpol General Secretary, Drug subdivision, Lyon.

Intraval (1987):

Integratie: mythe of werkelijkheid. De leefsituatie en drugsproblematiek van Surinamers, Antillianen en Arubanen in Breda nader bekeken. St. Intraval, Groningen.

Intraval (1989):

Harddrugs & criminaliteit in Rotterdam. St. Intraval, Groningen.

Intraval (1990):

Research design cocaine use in Rotterdam. St. Intraval, Groningen-Rotterdam.

Intraval (1991):

Door regelen in de maat. Reacties van drugsgebruikers op maatregelen door overheid en burger. St. Intraval, Groningen-Rotterdam.

Intraval (1992):

Between the lines. A study of the nature and extent of cocaine use in Rotterdam. St. Intraval, Groningen-Rotterdam.

Isner, J.M., S.K. Chokshi (1991):

Cardiac complications of cocaine abuse. Annual Review on Medicine, 42, 133-138.

Janssen, O., K. Swierstra (1982):

Heroïnegebruik in Nederland. Een typologie van levensstijlen. Criminologisch Instituut, RUG, Groningen.

Jeschke, K.H. (1987):

Measures to combat cocaine imports. The German viewpoint. In: Proceedings of the scientific meeting Cocaine. Office for the Official Publications of the European Communities, Luxembourg, 145-154.

Johanson, C.E., M.W. Fischman (1989):

The pharmacology of cocaine related to its abuse. Pharmacological Review, 41, 3-52.

Kaplan, Ch.D. (1984):
The uneasy consensus: prohibitionist and experimentalist expectancies behind the International Narcotics Control System. Tijdschrift voor Criminologie, 2, 98-109.

Kaplan, Ch.D. (1987):
The socio-economic consequences of drug abuse. In: Proceedings of the scientific meeting Cocaine. Office for the Official Publications of the European Communities, Luxembourg, 155-162.

Kaplan, Ch.D., D. Korf, C. Sterk (1987):
Temporal and social contexts of heroin-using populations. The Journal of Nervous and Mental Disease, 175, 9, 566-574.

Kaplan, Ch.D., B. Bieleman, W.D. TenHouten (1992):
Are there 'casual users' of cocaine? In: Cocaine: scientific and social dimensions. Wiley, Chichester (Ciba Foundation Symposium 166), 57-80.

Knapp, S., A.J. Mandell (1972):
Narcotic Drugs: effects on the serotonin biosynthetic systems of the brain. Science, 177, 1209-1211.

Knoke, D., J.H. Kublinski (1990):
Network Analysis. Sage Publications, Beverly Hills.

Korf, D.J., P.W.J. van Poppel (1986):
Heroïnetoerisme, veldonderzoek naar het gebruik van harddrugs onder buitenlanders in Amsterdam. Stadsdrukkerij van Amsterdam, Amsterdam.

Korf, D., M. de Kort (1989):
NV De witte waan: de geschiedenis van de Nederlandsche Cocaïnefabriek. NRC 13-5-89.

Korf, D., H. van Aalderen, H. Hoogenhout, P. Sandwijk (1990):
Gooise geneugten: legaal en illegaal drugsgebruik in de regio. SPCP, Amsterdam.

Kozel, N.J., E.H. Adams (eds.) (1985):
Cocaine use in America: epidemiologic and clinical perspectives. US Government Printing Office, Washington DC (NIDA research Monograph 61).

Kuhar, M.J., M.C. Ritz, J.W. Boja (1991):
The dopamine hypothesis of the reinforcing properties of cocaine. Trends Neuroscience, 14, 299-302.

Kuhar, M.J. (1992):
Molecular pharmacology of cocaine: a dopamine hypothesis and its implications. In: Cocaine: scientific and social dimensions. Wiley, Chichester (Ciba Foundation Symposium 166), 81-95.

Laumann, A., J.H. Gagnon, R.T. Michael, J.S. Coleman (1989):
Monitoring the AIDS Epidemic in the United States: a network approach. Science, 244, 1186-89.

Leroy, B. (1992):
The European Community of twelve and the drug demand. Excerpt of a comparative study of legislations and judicial practice. Drug and Alcohol Dependence, 29, 269-281.

Leuw, Ed. (1988):
Over gokken en de hernieuwde humanisering van het verslavingsbegrip. TADP, 14, 5-6, 178-185.

Leuw, Ed. (1991):
Drugs and drug policy in the Netherlands. Dutch penal law and policy 04, 11, 1991.

Lewis, R. (1989):
European markets in cocaine. Contemp. Crises, 13, 35-52.

Limbeek, J. van (ed.) (1986):
Cocaïne. FZA, Bilthoven.

López-Muñiz, M. (1980):
Tratamiento legal de la drogodependencia en España. In: Actas del IX Congreso Internacional sobre prevención y tratamiento de las drogodepencias. INSERSO, Madrid, 159-171.

Macchia, T., R. Mancinelli, G. Bartolomucci, U. Avico (1990):
Cocaine misuse in selected areas. Ann. Ist. Sup. Sanità, 26, 2, 189-196.

Malinowski, B. (1967):
A diary in the strict sense of the term. Harcourt, Brace and World, New York.

Mayer, A.C. (1966):
The significance of quasi-groups in the study of complex societies. In: M. Banton (ed.): The social anthropology of complex societies. Travistock Publications, London, 97-122.

McBride, D. (1981):
Drugs and violence. In: J.A. Inciardi (ed.): The drugs-crime connection. Sage Publications, Beverley Hills, 105-123.

Medina-Mora, M.E., P. Ryan, A. Ortiz, T. Campos, A. Solís (1980):
Metodología para la identificación de casos y la vigilancia del uso de drogas en una comunidad mexicana. Boletín de estupefacientes, 32, 2, 19-29.

Meerten, R. van (1992):
Cocaïne, de bijsluiter (concept). Boumanhuis, Rotterdam.

Merlo, G. (1990):
Il sistema di osservazione nella città di Torino. In relazione al Convegno La droga: osservare il fenomeno organizzare i servizi, 26-1-1990. ISSOS, Napoli.

Merlo, G., F. Borazzo, U. Moreggia, M.G. Terzi (1992):
Network of powder. Research report on the cocaine use in Turin. Ufficio Coordinamento degli Interventi per le Tossicodipendenze, Torino.

Meulen, J.D. van der (1969):
Drugs en het strafrecht. In: C. Wijbenga (ed.): Soft Drugs. Van Gennep, Amsterdam.

Mitchell, J.C. (1966):
Theoretical orientations in African urban studies. In: M. Banton (ed.): The social anthropology of complex cities. Tavisstock Publications, London, 37-68.

Morningstar, P.C., D.D. Chitwood (1983):
The patterns of cocaine use: an interdisciplinary study. Final report no. R01 DA03106. NIDA, Rockville MD.

Morningstar, P.C., D.D. Chitwood (1984):
Cocaine users view of themselves. Implicit behavior theory in context. Human Organization, 43, 307-318.

Morningstar, P.J., D.D. Chitwood (1987):
How women and men get cocaine: sex-role stereotypes and acquisition patterns. J. Psychoactive Drugs, 19, 2, 135-142.

Murphy, S.B., C. Reinarman, D. Waldorf (1989):
An 11-Year follow-up of a network of cocaine users. Brit. Jour. Addiction, 84, 427-436.

Musto, D.F. (1973):
The American decease. Yale University Press, London.

Musto, D.F. (1992):
Cocaine's history, especially the American experience. In: Cocaine: scientific and social dimensions. Wiley, Chichester (Ciba Foundation Symposium 166), 7-19.

NDSN (1992):
National Drug Strategy Network. Newsbrief, 3, 7, 7.

NIDA (1986):
Prevention Networks: Cocaine use in America. NIDA, Washington DC.

NIDA (1990):
Population estimates of lifetime and current drug use. NIDA, Washington DC.

NVC (1990):
Ladis 1988: jaarstatistieken uit het landelijk alcohol en drugs informatie-systeem. NVC, Utrecht.

NVC (1992):
Ladis 1991: jaarstatistieken uit het landelijk alcohol en drugs informatie-systeem. NVC, Utrecht.

Organ Tècnic de Drogodependències (1992):
Informació de l'Organ Tècnic. Departament de Sanitat de la Generalitat de Catalunya, Barcelona.

Oro, S., S.D. Dixon (1987):
Perinatal cocaine and methamphetamine exposure. Maternal and neonatal correlations. Journal of Paediatrics, 111, 571-578.

Peele, S. (1987):
A moral vision of addiction: how people's values determine whether they become and remain addicts. J. Drug Issues, 17, 1-2, 187-215.

Petersen, R.C., S. Cohen, F.R. Jeri, D.E. Smith, L.I. Dogoloff (1983):
Cocaine: a second look. The Am. Council on Marijuana and Other Psychoactive Drugs, New York.

Phillips, J., B. Frost (1992):
Cocaine ring smashed in seven million pound seizure by international agencies. The Times, 29-9-1992.

Plan Nacional Sobre Drogas (PNSD) (1992):
Memoria 1991. Plan Nacional Sobre Drogas, Madrid.

Plomp, H.N., H. Kuipers, M.L. van Oers (1990):
Roken, alcohol- en drugsgebruik onder scholieren vanaf 10 jaar. Resultaten van het vierde peilstationsonderzoek jeugdgezondheidszorg 1988/1989. VU Uitgeverij, Amsterdam.

Pompidou Group (1986):
Multi-city report. European Council, Strasbourg.

Pompidou Group (1992):
Activities in the field of epidemiology. Paper presented at the Symposium: A European Drug Monitoring Centre Health Related Data and Epidemiology in the European Community. Brussels, 1992, september 21-23.

Ponti, G., E. Calvanese, F. Marozzi, I. Merzagora. (1991):
Cocaina en criminalità: rapporto diretto o rapporto d'am biente? In: Atti del seminario Cocaina oggi: Effetti sull'uomo e sulla società. UNICRI, Roma, 281-290.

Post, R.M. (1975):
Cocaine psychosis: a continuum model. The American Journal of Psychiatry, 132, 225-231.

Putten, J. van der (1992):
V.S. en Italië rollen enorme organisatie op van drugsmafia. De Volkskrant, 29-9-1992.

Ree, F. van, P. Esseveld (1985):
Drugs: de medische en maatschappelijke aspecten. Spectrum, Utrecht/Antwerpen.

Rodríguez, B. (1989):
Evolución de la tendencia de mortalidad por reacción aguda tras el consumo de drogas en Madrid, Valencia y Barcelona. Período 1983-1989. Delegación del Gobierno para el Plan Nacional sobre Drogas, Madrid.

Romaní, O., N. Espinal, J.M. Roviro (1989):
Presa de contactes ambels drogodependents d'alt risc. Ajuntament de Barcelona, Barcelona.

Romaní, O,, C. Rimbau, N. Espinal, J. Pallares, R. Romero, J. Cañellas (1991):
Drogodependientes, circuitos informales y procesos de integracion social. IRES, Barcelona.

Rook, A., J. Essers (1987):
Vervolging en Strafvordering bij Opiumwetdelicten. WODC nr. 80. Staatsuitgeverij, 's-Gravenhage.

Rouse, B.A. (1991):
Trends in cocaine use in the general population. In: S. Schober, C. Schade (eds.): The epidemiology of cocaine use and abuse. US Government Printing Office, Washington DC (NIDA research Monograph 110), 45-56.

Sandwijk, J., I. Westerterp, S. Musterd (1988):
Het gebruik van legale en illegale drugs in Amsterdam: verslag van een prevalentie-onderzoek onder de bevolking van 12 jaar en ouder. Inst. Soc. Geogr., UvA, Amsterdam.

Sandwijk, J.P., P.D.A. Cohen, S. Musterd (1991):
Licit and illicit drug use in Amsterdam. Inst. Soc. Geogr., UvA, Amsterdam.

Schatzman, L., A.L. Strauss (1973):
Field research: Strategies for a natural sociology. Prentice-Hall, New Jersey.

Schein, E.H. (1987):
The clinical perspective in fieldwork. Sage Publications, London.

Schuster, C.R., M. Fischman, C.E. Johanson (1981):
Internal stimulus control and subjective effects of drugs. In: C.E. Johanson, T. Thompson (eds.): Behavioral Pharmacology of human drug dependence. US Government Printing Office, Washington DC (NIDA Research Monograph 37), 116-129.

Scott, J. (1991):
Social network analysis: methods and applications. To be published by Cambridge University, New York/Cambridge.

SIDB (1991):
Sistema D'Informació de Drogodependències de Barcelona. Informe anual corresponent a l'any 1990. Institut Municipal de la Salut, Barcelona.

SIDC (1992):
Sistema d'Informació sobre Drogodependències a Catalunya. Informe anual 1991. Organ Tècnic de Drogodependències, Barcelona.

Siegel, R.K. (1984):
Changing patterns of cocaine use: longitudinal observations, consequences, and treatment. US Government Printing Office, Washington DC (NIDA Research Monograph 50).

Siegel, R.K. (1985):
Treatment of cocaine abuse: historical and contemporary perspectives. J. Psychoactive Drugs, 17, 1, 19851-9.
Sistema Estatal de Información sobre Toxicomanías (SEIT) (1988):
Informe 1987. Plan Nacional Sobre Drogas, Madrid.
Sistema Estatal de Información sobre Toxicomanías (SEIT) (1989):
Informe 1988. Plan Nacional Sobre Drogas, Madrid.
Sistema Estatal de Información sobre Toxicomanías (SEIT) (1990):
Informe 1989. Plan Nacional Sobre Drogas, Madrid.
Sistema Estatal de Información sobre Toxicomanías (SEIT) (1991):
Informe 1990. Plan Nacional Sobre Drogas, Madrid.
Snijders, T.A.B. (1991):
Estimating the size of a network. University of Groningen, Groningen.
Snijders, T.A.B. (1992):
Estimating on the basis of snowball samples: how to weight? Bulletin de Methodologie Sociologique, 36, 59-70.
Spotts, J.V., F.C. Shontz (1980):
Cocaine users: a representative case approach. Free Press, New York.
Spotts, J.V., F.C. Shontz (1984):
Drug-induced ego states. Int. J. Addict., 19, 2, 119-151.
Spreen, M. (1992a):
Rare populations, hidden populations, and link-tracing designs: what and why? Bulletin de Methodologie Sociologique, 36, 34-58.
Spreen, M. (1992b):
Star sampling, out-degree analysis and multilevel analysis. A practical link-tracing methodology for sampling and analyzing sociocentric hidden populations. Department of Sociology, University of Groningen, Groningen.
Stein, S.D. (1985):
International diplomacy, state administrators and narcotic control: the origins of a social problem. Gower publ. company lim., Hampshire.
Stone, N., M. Fromme, D. Kagan (1984):
Cocaine: seduction and solution. Potter, New York.
Sudman, S., M.G. Sirken, C.D. Cowan (1988):
Sampling rare and elusive populations. Science, 24, 20, 991-96.
Swierstra, K., O. Janssen (1986):
De reproductie van het heroïnegebruik onder nieuwe lichtingen. Heroïnegebruik in Nederland deel II. Criminologisch Instituut, RUG, Groningen.
Swierstra, K. (1990):
Drugscarrières. Van crimineel tot conventioneel. Onderzoekscentrum voor criminologie en jeugdcriminologie, RUG, Groningen.
Taradash, M. (1991):
Legalizzare la droga. Una ragionevole proposta du sperimentazione. Feltrinelli, Milano.

Tardiff, K., E, Gross, J. Wu, M. Stajic, R. Millman (1989):
From victims to criminals to victims. In: J.A. Inciardi (ed.): The drugs-crime
63.
Tieman, C.R. (1981):
From victims to criminals to victims. In: J.A. Inciardi (ed.): The drugs-crime
connection. Sage Publications, Beverley Hills, 239-267.
Toet, J., R. Geurs (1992):
Resultaten RODIS 1991; concept (versie 2). GGD Rotterdam e.o., afdeling
epidemiologie/OGGZ, Rotterdam.
Van Dyke, C., R. Byck (1982):
Cocaine. Scientific American, 246, 3, 128-142.
Van Meter, K.M. (1990):
Methodologies and design issues: Techniques for assessing the
representatives of snowball samples. In: E. Y. Lambert (ed.): The collection
and interpretation of data from hidden populations. US Government Printing
Office, Washington DC (NIDA Research Monograph 98), 31-43.
Volpe, P.P. (1992):
Effect of cocaine use on the fetus. The New England Journal of Medicine,
327, 6, 399-407.
Waldorf, D., C. Reinarman, S. Murphy (1991):
Cocaine changes. The experience of using and quitting. Temple University
Press, Philadelphia.
Wasserman, S., K. Faust (in press):
Social network analysis: Methods and applications. To be published by
Cambridge University Press, New York/Cambridge.
Watters, J.K., C. Reinarman, J. Fagan (1985):
Causality, context, and contingency: relationships between drug abuse and
delinquency. Contemp. Drug Problems, 1985, 351-373.
Watters, J.K., P. Biernacki (1989):
Targeted sampling: options for the study of hidden populations. Social
problems, 36, 4, 416-430.
Wax, R. (1971):
Doing fieldwork. University of Chicago Press, Chicago.
Werner, O. (1988):
An anthropological approach to fieldwork: Ethnoscience. In: L. Borzak (ed.):
Field study: A sourcebook for experimental learning. Sage Publications,
London.
Wheeldon, P.D. (1969):
The operation of voluntary associations and personal networks in the
political processes of an inter-ethnic community. In: C. Mitchell (Comp.):
Social Networks in Urban Situations. University of Manchester Press,
Manchester, 128-180.

Wijngaart, G.V. van de (1991):

Competing perspectives on drug use. The Dutch Experience. Swets & Zeitlinger, Amsterdam/Lisse.

Williams, T. (1989):

The cocaine kids: the inside story of a teenage drug ring. Addison-Wesley, New York.

Wise, R.W. (1984):

Neural mechanisms of the reinforcing action of cocaine. In: J. Grabowski (ed.): Cocaine: pharmacology, effects and treatment of abuse. US Government Printing Office, Washington DC (NIDA Research Monograph 50).

WVC (1992):

The drug abuse situation in the Netherlands. Ministry of Welfare, Health and Cultural Affairs & Ministry of Justice, Rijswijk.

Zinberg, N.E. (1984):

Drug, set, and setting: the basis for controlled intoxicant use. Yale University Press, New Haven/London.

Zuidhof, F.C. (1984):

125 jaar cocaine. TADP, 10, 3, 129-133.

GLOSSARY

Basing
Method of use. Cocaine alkaloid (crystals) are smoked in a pipe, often filled with a strong alcoholic drink, for instance rum. Sometimes a glass and foil or silver paper are used.

A Bomb
Speed or cocaine is wrapped in a piece of paper and then swallowed.

Bouncing
A negative effect of cocaine use. People feel very fed up, so they can not relax and are unable to sleep. This is often ascribed to lacing with speed.

Brown
Name used for heroin, especially popular among hard drug users (see also White).

Chasing the dragon
Method of use. Cocaine is heated on foil or silver paper and the fumes are inhaled through a tube.

Coke
Most frequently used name for cocaine. Sometimes more lyric descriptions such as snow, the king of drugs, powdered sugar or Columbian marching powder are used.

Coke blow
Name for a cigarette or fag containing cocaine. Also called a coke joint.

Cooked coke
Cocaine which is boiled with the help of sodium bicarbonate or ammonia. The boiled cocaine can be used by basing or chasing the dragon. It is uncertain if this has to be considered the same as crack. Several users call it a Dutch variation of crack.

Crack	The hydrochloride component of cocaine, distilled with the aid of soda and water to a 'base' (with free-basing this is done using volatile substances). It can be smoked, for example in a pipe.
A Line	A varying amount of cocaine, depending on the length and thickness of the line.
A Run	Continuous cocaine use during, for example, one or two days.
Shotting	Method of intravenous use of cocaine. Cocaine is injected with a syringe. This method is popular only among poly-drug users.
Smack	Name for heroin. Used especially by poly-drug users.
Smoking	Method of use. Cocaine is put in a cigarette or fag and smoked. It can be compared with smoking cannabis. This is considered an expensive method, because a lot of the active substances are lost.
Sniffing/Snorting	Most common way of using cocaine. With the aid of a tube the cocaine is inhaled through the nose.
White	Name for cocaine. Used especially by poly-drug users (in contrast to brown: heroin).

APPENDICES

APPENDIX A

A MARRIAGE OF POLITICS AND SCIENCE

A. Arias, G. Merlo, R. van der Steen

The gestation of Convention No 91CVVE1259-0 'Coordination Europeén-ne d'une analyse de l' Usage de la Cocaïne dans les Villes de Rotterdam, Barcelone et Turin'.

The municipalities of Barcelona, Rotterdam and Turin and the General Directorate V of the European Commission allocated the funds that made the study possible. Since the project was originally intended to include several European centres, it was extremely important to have the support and sponsorship of the Commission of the European Community. The individual cities would not have been able to achieve their aim without the European Commission's moral and financial support. This project was unique in terms of cooperation between political administrators, policy planners and researchers.

Background of the project
Several coincidental factors created a feeling of anxiety about cocaine at the European level. In the late Eighties this concern led to the gestation of the project. Three factors should be mentioned.
- Mass-media information on the increasing entry of cocaine into Europe, mainly through Spain and the Netherlands, taken together with the sensationalized crack epidemic in the USA.
- The social perception of widespread use of cocaine among several prominent social circles such as entertainment and nightlife celebrities.
- An increasing trend in the indicators of the adverse health effects of cocaine (mainly in hospital emergencies related to cocaine) as well as in cocaine seizures detected across Europe.

Two main premises underlied the agreement to develop this project. First, there was the feeling that any approach to address the cocaine issue (as with all drug issues) needs to be considered transnationally since the supply characteristics and the cultural conditions of demand go beyond the borders of an individual country. Second, there was the need for more information on the characteristics of cocaine use. This knowledge was given a even a higher priority than the knowledge of the extent of use. This led to two main questions.

- With the large increase in seizures, why were there no observable parallel increases in social and health adverse reactions to cocaine?
- Do cocaine use patterns in Europe differ from those in the USA?

Basic facts

In 1989, during the 2nd International Conference on the Reduction of Drug Related Harm in Barcelona the political decision makers of social and public health affairs of the cities of Rotterdam and Barcelona made an initial agreement to develop a project to find out more about cocaine use in those cities. Since the municipality of Rotterdam made the initial proposal, it was agreed that Rotterdam would lead the project in terms of basic technical and scientific issues, coordination of work and the international projection of the research. Based on project premises, there was an awareness of the need to include more European centres. The cities of Turin and Lyon were invited to participate. Turin found the proposal interesting, opportune and timely. Lyon decided not to participate in the project.

During discussions among the policy planners and political decision makers of the three cities, specific basic requirements of the project emerged. Thus, the project had to:

- address current worries and concerns about cocaine use;
- be able to anticipate future problematic cocaine issues;
- be both scientifically rigorous and useful for policy development;
- allow not only for validity at city level, but also for comparability among the three cities.

Organization of the project

The City Council of Rotterdam, represented by Mr. Henderson, commissioned the private foundation Intraval, an agency for social scientific research and consultancy, to execute the study. Mr. Bieleman, the Executive Director, was assisted by Mrs. Ten Den, Mr. De Bie and Mr. Spreen. Mr. Henderson delegated the coordination concerning the international contacts and the overall follow-up to Mrs. Verlaan and the overall control of financial matters to Mr. Van der Steen. They were also put in charge of the policy preparation on the implications of the study.

Although based on the Rotterdam ideas and research design proposed by Intraval, there were slight differences in the way each city organized the actual study. The City Council of Barcelona, first represented by Mr. Clos and later by Mr. Casas, chose to entrust the research to the Sociology Laboratory of the University (ICESB). The main group of researchers consisted of the Director Mr. Díaz assisted by Mrs. Barruti, Mrs. Doncel and Mr. Puig. Mr. Arias was responsible for following-up the process, the international contacts and the development of policy preparation.

The City Council of Turin, first represented by Mr. Braco and later by Mr. Simonetta, decided that the research should be done by the Municipal Health Unit's central agency coordinating intervention in drug addiction. Mr. Merlo was the main person responsible, assisted by Mrs. Terzi, Mr. Borazzo and Mr. Moreggia. Mr. Merlo was also in charge of the international contacts and policy preparation.

As the project started, many other scientists became involved including field workers in each participating city and the expert advisers Prof. Del Castillo of the Autonoma University of Barcelona and Prof. Snijders of the University of Groningen who were asked to develop new methods for estimating the numbers of users. Finally, the three cities agreed to invite Professor Kaplan to coordinate the international report. In the context of the developing project, Professor Snijders organized a two day workshop in Groningen, in February 1992. The Workshop discussed the generalizability of questions for snowball sampling and other ascending methodologies. A listing of the many other participants in the project is presented in Appendix B.

Development of the project

In order to achieve the aims of the project, participation at three functional levels was required: the research teams, the municipal policy planners, and the municipal decision makers. The decision makers defined the general focus of interest of the project for the policy planners. They periodically assessed and monitored the results, conclusions, and recommendations of the researchers. The policy planners defined the objectives of the project and selected the research teams. The implementation of the project included the following tasks: coordination of structural and functional needs of project development; maintenance of periodic contact with the research teams; preparation of follow-up meetings in the three cities and fitting the project development to the objectives and priorities defined by the three city councils. The research teams had the task of designing the project, developing the fieldwork, making the analysis and reporting. They also had to organize the exchange of information between the three cities and revise their methodological approaches in order to ensure maximum uniformity.

The follow-up of the international project and discussions among the three cities were mainly based on three, four day meetings convened sequentially in each city. Progress reports by the research teams were discussed at the meetings. The first days were devoted to scientific discussions and an exchange of information on the development of the local projects. The aim was to update information and homogenize the methodological and analytical approaches. Political decision makers, policy planners and research teams joined together on the last day of the meeting. The work done by the research teams on the preceding days was presented and the provisional results and conclusions discussed.

Developments relating to the European Commission

Although, in April 1991, the actual cooperation between the cities was initiated with a first working meeting in Turin, it was uncertain whether the European Commission would be willing or able to contribute funds to the study. The Rotterdam City Council was delegated by the three cities to make the formal application for funds from the Commission. The completed application forms were sent to Luxembourg at the end of June 1991. Subsequently, the European Commission released funds to a total amount of 180,000 ECUs. This grant covered about two third of the estimated expenses of the project starting at the end of December 1991. At the same time, the Commission selected Rotterdam City Council as the coordinating body. This meant that the city would take full responsibility for the study in terms of deadlines, administration and overall financial control.

For administrative reasons, the European Commission closes its financial year on 31st October. This made it necessary to complete the comparative research study and the final financial account within ten months, rather than in the twelve that had been originally planned. This unforeseen time limit significantly increased work pressure on the researchers, decision makers and policy planners. A great deal of self-discipline and an increased tempo of communication between the different functional groups was required and achieved. Through this special effort, outstanding results were produced. The end conclusion is that it is indeed possible, necessary, and important to have ongoing cooperation among European cities on matters relating to drug policy development and research.

APPENDIX B

ACKNOWLEDGEMENTS

Many engaged people in various functions devoted time and energy to this project. We cannot thank them all and we apologise for those we have left out. Nevertheless, without the following people, the project could not have been realized as it was.

Barcelona

City Council/Politicians:
  Mr. Clos, Mr. Casas
City Hall:
  Mr. Arias
Research institute:
  Sociology Laboratory of the ICESB
Project manager:
  Aurelio Díaz
Coordinator, systemise information and analysis:
  Concha Doncel
Qualitative analysis and leader of field work team:
  Mila Barruti
Estimations:
  Joan del Castillo
Statistics:
  Pedro Puig
Data processing, analysis and presentation:
  Conchita Díaz, Maria Recolons
Field work:
  Mila Barruti, Marisa Clavero, Judit Domènec, Concha Doncel, Silvia Donoso, Núria Espinal, Mireia Ferré, Ricard Ginés, Antonia Lloret, Jordi

Moreras, Manel Nadal, Pau Nadal, Jordi Noya, Núria Pahissa, Nela Rodríguez, Glòria Rosset, Puri Ruiz, Yasmin Seyed
Advisors:
   Antoni Arias, Joan Maria Pinyol, Joan Josep Pujadas, Santiago Perera, Antoni Ruiz, Oriol Romani, Lluis Torralba
Catalan:
   Ramon Cotrina
English translation:
   Teacher's Group
Assistance:
   Tre Borràs, Eugeni Bruguera, Jordi del Rey, Jaume Funes, Ana Gil, Antonio Lage, Núria Magrí, Francesca Mata, Manuel Más-Bagá, Ernesta Sánchez, Luis de la Fuente, Elisenda Giralt, José Maria Mena and Julio Zino, Carmen Guilayn, Anna Pomares, Maja Stivilj
Collaboration:
   All the professionals in the different institution and services, both private and public, to which we addressed ourselves during the different phases of the project.
Thanks:
   All the persons who have been interviewed, those who performed as go-betweens in these interviews so that they could be carried out and those who assisted us in highly different ways during the field work. To all of them, our most singular gratitude.

Rotterdam

City Council/Politician:
   Mr. Henderson
City Hall:
   Mrs. Verlaan, Mr. Van der Steen
Research institute:
   INTRAVAL, bureau for social-scientific research and consultancy
Project manager:
   Bert Bieleman
Qualitative analysis and leader of field work team:
   Edgar de Bie
Characteristics:
   Cilia ten Den
Network analysis and estimations:
   Marinus Spreen
Field work:
   Edgar de Bie, Otto Doosje, Ageeth Ettema, Lucas Kroes, Jan Vis

Scientific advisors:
    Prof. Snijders (network analysis, estimations and methodology), Prof. Kaplan (drug use and abuse), Dr. Brandsma (research design and methodology)
Advisors:
    Jolt Bosma, Paula Koedijk, Eddy Leuw, Koert Swierstra
Secretariat:
    Nicolien de Vries, Jacco Snippe, Reiko Shiota
English translation:
    Mary Marggraf-Lavery of Maarn Translations, Jim Allen
Assistance field work:
    Bas Berkhout, Adèle Hoekstra, Rob van Meerten, Andras Simon
Collaboration:
    Auke de Jong, steering committee, the Rotterdam City Police, staff of the Bulldog and other (drug) assistance agencies
Thanks:
    The cocaine users themselves who discussed their experiences with us in lengthy interviews. Without their frank and enthusiastic cooperation this study would not have been possible.

Turin

City Council/Politician:
    Mr. Bracco, Mr. Simonetta
City Hall:
    Mr. Merlo, Mrs. Chianale
Research institute:
    Central Bureau Coordinating the Intervention for Drug Addiction of the Local Health Unit
Project manager:
    Giorgio Merlo
Analysis and leader field work team:
    Maria Grazia Terzi
Statistics:
    Francesco Borazzo
Sociological analysis:
    Uberto Moreggia
Field work:
    Vittorio Castellani, Silvana Dutto, Giuliano Mochi Sismondi, Sara Negarville, Gabriele Pagella, Rossana Rosso
Analysis of daily newspapers:
    Giuliano Mochi Sismondi

Secretariat:

Vera Gambini

English translation:

Rita Garriba

Collaboration:

Prefecture of Turin, Giuseppe Forlani; Police Force, Pasquale Muggeo, Roberto Bovi, Francesco De Cicco; Chemical Laboratory of USSL 1, Franca Ricottilli, Gian Piero Spagnolini

Assistance:

The Health Centres for Drug Action of the Local Health Units of Turin, Centro Torinese de Solidarieta, Cooperativa ESEDRA

Thanks:

All the interviewees for making this work possible.

APPENDIX C

SAMPLING AND ANALYZING IN HIDDEN POPULATIONS

Marinus Spreen[1]

Data collection procedures using contact patterns between members of hidden populations[2] are frequently applied to obtain a reasonable amount of respondents in studies of health and social problems. Examples of such populations are heroin addicts, cocaine users, hooligans, street gangs, criminal organizations, and illegal aliens. Contact patterns are used for sampling purposes because conventional sampling methods are an inefficient means for reaching a statistically sufficient proportion of the target population. Achieving a proportion sufficient to make valid statistical inferences is problematic because hidden populations are usually a very small proportion of the total population at large, sampling frames are frequently absent, the individuals are hard to locate, or to make contact with, and it is often difficult to determine whether a given individual actually belongs to the target population. Getting reliable information about the respondents is problematic, too; it is necessary to use intensive interviewing techniques. All these problems contribute to the fact that applying standard probability samples in studies of hidden populations is mostly impractical or impossible, which means that valid statistical inferences cannot be made. Given these difficulties it is hardly surprising that most studies of hidden populations continue to be qualitatively oriented.

The fact that contact patterns can be applied as a sampling tool is an indication that the variable which defines the hidden population has a social meaning in the sense that members know other members precisely because of the nature of this variable. For instance, cocaine users are supposed to have bonds with other cocaine users because cocaine is considered to be a social drug. The structure of the personal networks as well as the entire network will influence the behaviour of individuals, for instance with regard to the opportunity an individual has to use or obtain cocaine. Therefore, in studies of hidden populations it is worthwhile to take network structure into account, not only in the data collection but also in the analysis of the population. In 1958 Coleman argued that in studies of social groups or behaviour, sampling designs must involve a strategy to collect data pertaining to the respondent's relation to other specific individuals. This perspective has been developed in social network analysis. Most traditional network methods and parameters, developed to capture influences of network structure on the behaviour of actors and institutional arrangements, and vice versa, are appropriate only for situations in which the entire network is available as observational data. Consequently, in social network analysis the most common sampling designs are some form of saturation samples (all members observed or assumed to be observed) and as a result the elaborated network parameters are restricted to the analysis of small and bounded groups.

The sample techniques using contact patterns reported in most literature are called 'snowbally' methods, and mainly regarded and applied as what Kruskal and Mosteller (1979, 1980) define as chunk methods[3]. However, it is worthwhile to apply these methods in a more formal way for making well-founded inferences about important aspects of network structure. Therefore, we prefer to define these sample techniques as link-tracing designs because:

- a snowball procedure is often considered to be a self-contained and self-propelled phenomenon, in that once it is started it somehow magically proceeds on its own (Biernacki and Waldorf 1981);
- snowball sampling is part of a family of sampling techniques the purpose of which is to sample explicitly with reference to network structure, i. e., to trace links in a large network.

For a review of link-tracing designs see Spreen (1992a).

Hidden populations will often be a minor proportion of the population at large, but this does not imply that this population can be regarded as a small and bounded group. For instance, the estimated number of cocaine users in Barcelona and Rotterdam is in the range of 5,000 - 35,000 (see chapter 7), which means that a saturation sample of the total group is hardly feasible. Additionally, even if the hidden population consists of only 100 persons, a saturation sample would be also infeasible because it is quite likely that only a portion of the population could be identified and would be willing to co-operate. At best, only a part of the network of interest can be sampled and attempts at describing network structure in hidden populations can be made only from a fragment of the population. This appendix introduces and evaluates a practical link-tracing methodology for analyzing hidden populations (as applied in the cocaine research of Rotterdam): star sampling, out-degree analysis, and multilevel analysis. It is suggested, that by means of this methodology, well-founded inferences can be made about at least two important aspects of the network structure of cocaine users: the opportunity structure as well as the significance of cocaine for the relationships in connection with the opportunity structure.

Methodological aspects of sampling designs in studies of hidden populations

In studies of hidden populations emphasis is primarily placed on the representativeness of the collected sampling units with respect to the whole target population. According to Kish (1987) the purpose of a representative sample is to make the sample a miniature so as to mirror and to represent the population with similar distributions; such representation serves the purpose of the intended inferences. This can only be achieved by using some kind of probability mechanism for drawing members of the population, so that for each individual the probability of being in the sample can be estimated and selection bias avoided. Hidden populations are characterized by the impracticability of applying probability samples, so statistical inferences cannot be made. However, by using a non-probability design one can try to approximate as closely as possible a probabilistic representative sample. This creates the possibility to make well-founded inferences.

The best way of approximating a random sample is to collect respondents as 'physically' independently of each other as possible (Snijders 1992). For instance, when in some hidden population members are expected to be found in places of entertainment, only one respondent per place of entertainment should be used in the analysis to approximate randomness in the nonprobability sample. Naturally, within each place of entertainment, or more generally formulated small-scale 'social environments', more respondents might be found and interviewed (Snijders 1992). When more respondents per small-scale social environment are collected, these respondents should only be added to the 'independently selected' sample if the researcher (or fieldworker) can verify that these respondents are found as much as possible independently of each other. This implies that the time, place, and way in which each interviewed respondent is found, must be known. Therefore, in order to make well-founded inferences when sampling members of a hidden population by means of a non-probability selection method, it is necesssary to have a detailed methodological logbook detailing the independence of finding respondents.

When making well-founded inferences about the structure of a network, the inclusion criterion that defines the boundaries of the network is essential. This inclusion criterion must define unequivocally which persons and relations constitute the network of interest. According to Marsden (1990) the most central question for network measurement is the extent to which the data collection method is able to identify social contacts. Some important issues of network measurement must be taken into consideration when defining inclusion criteria. Inclusion criteria must make explicit the kind of relations to be measured: for instance social relations which are long-lasting or social relations which are defined by occurrences in very short time frames. Another issue concerning network measurement is that of informant accuracy; does the respondent mention all his relations concerning the inclusion criterion? Particularly in network studies of hidden populations these issues deserve special attention because:

1. Respondents will be reluctant to mention other people. If they do, what kind of relations do they mention? Only good friends or relative strangers or both?

2. Respondents will be reluctant to disclose issues concerning the variable that defines the network.

There are no ways of avoiding this kind of bias but in the case of in-depth interviews in which the interviewer has to gain the trust of the respondent, one has to assume that respondents will give accurate information about their relations.

Another type of selection bias we must reckon with in a sampling design for hidden populations is the inevitable over-representation of the more socially active persons in the sample. These persons are simply more likely to be found by the fieldworkers. By assuming that the more socially active members will be known to many other members in the network (i.e. have a high in-degree[4]), a control variable should be included into the research design to obtain at least a rough indication of the in-degrees by posing questions such as: 'What is your estimate of the number of persons in the network who would say they have the relation (according to the inclusion criterion) to you?'. Another way of getting an indication of the in-degrees is to define the inclusion criterion symmetrically by assuming that each respondent mentions relations who, in turn, would also mention this respondent, but this precludes the study of nonsymmetrical relations (in that case the researcher has to know the out-degree, see note 4).

When using a link-tracing design as a more or less formal sampling method, four purposes can be distinguished (Snijders 1992):

1. making inferences about the population of individuals;
2. making inferences about the population of relations;
3. making inferences about the population of personal networks;
4. making inferences about the total network.

Diagram 1 provides an overview of various link-tracing designs and their purposes.

Diagram 1   Overview of various link-tracing designs and their purposes

Inference about	Link-tracing design
Individuals	Multiplicity Sampling (Sirken 1970)
Relations	Snowball Sampling (Goodman 1961, Frank 1977)
Personal networks	Star Sampling (Frank 1971, Capobianco & Frank 1982)
Total network	Snowball Sampling (Goodman 1961, Frank 1977) Random Walk Approach (Klovdahl 1989)

When making inferences about individuals in a network, one must reckon with the fact that the probabilities of selecting individuals for extension of the sample is dependent on the initial sample. The most usual method of deriving estimators for parameters such as population averages is in this case the Horvitz-Thompson method (Horvitz and Thompson 1952). However, this method needs knowledge of the inclusion probabilities of all sampled individuals. These inclusion probabilities can be computed as functions of the network structure, but as Snijders (1992) states: 'This is nice in theory, but how can these inclusion probabilities be observed? You can compute them, in principle, but observe or estimate ..?' Snijders shows that for computing the inclusion probability of the vertices in the first extension, at the very least, the in-degrees of the respondents must be known. This amounts to asking: 'How many individuals would say you are their friend.' Since it is impossible to obtain a reliable answers to this question, in-degrees are often unobservable. An exception is multiplicity sampling, in which it is possible to give the in-degrees of the nominees. This is possible because the variable that defines the rare population does not provoke its members to be unwilling to disclose information. Snijders (1992) concludes that when using link-tracing designs for purpose 1, it is best to use only symmetric relations and a low number of extensions.

Link-tracing designs are better suited for making inferences about relations or about the total network. Goodman (1961) and Frank (1977) estimate several kind of network parameters for purposes 2 and 4. Klovdahl (1989), with his random-walk design, aims to describe connectedness in urban social networks.

These designs are not discussed in this appendix, because they have common assumption of a relatively easy access to the members of the network and an easy way of extending the sample, so that a probability sample is feasible. It is precisely these assumptions that distinguish hidden populations from other populations.

Another way of capturing network structure is by means of personal networks. Important aspects of network structure can be identified by personal networks. The properties of the personal network can be aggregated to the respondent, which technically transforms purpose 3 to an instance of purpose 1. This perspective provides a practical way of capturing aspects of network structure in studies of hidden populations.

Star sampling: sampling personal networks

Sampling a personal network means that for each sampled respondent, all his adjacent nominees (or relations) are observed. Capobianco and Frank (1982) discuss star sampling (i.e. sampling personal networks), but they assume knowledge of the identities of the respondents as well as the nominees, so the number of relations can be observed between sampled individuals and also between sampled and non-sampled individuals. In the model considered in this appendix, which is elaborated for hidden populations, it is not assumed that the nominees in the personal networks need to be observed (disclosed). The model has the following features:

1. It is a one extension link-tracing model for a given, fixed but a priori unknown, network of size N (N also is a priori unknown). It is assumed that the population of individuals as well as the relations among them are unequivocally defined.
2. The individuals in the network are assumed to have a number of relations that may vary from individual to individual.
3. A simple random sample of n individuals is drawn from the network. For the respondents in the initial sample their relations are observed.

This simple model has several important practical implications in studies of hidden populations. The respondent need not identify nominees for extending the sample, so he may be more willing to give (reliable) information about others. Another practical advantage is that this model avoids usual bias in identifying the nominees (who is who). The star sampling design will also avoid chaining processes and masking (Erickson 1978). Chaining processes occur whenever people have some choice in generating the traced chains so that these chains are not fully determined by the network structure. One reason why chaining processes occur, is the phenomenon of masking. Masking occurs when respondents are willing enough to report their relationships but are unable to do so accurately because they have more, or fewer, relationships than they are asked to report. Another practical advantage is that the model can be applied in a multiple extension link-tracing design because it is simply the first stage.

In most studies of hidden populations multiple extension link-tracing designs will be used because of the impossibility of achieving a fair amount of initial respondents. By also taking the star sample into account during the analysis, different samples can be compared in order to examine possible differences between them which might be due to selection processes (like chaining processes, masking, a sample weighted too heavily in the centre of the network etc.) in the total sample. In this way it is possible to make well-founded inferences about some important aspects of the network structure when using a multiple link-tracing design. Two methods to analyze data obtained by star sampling are out-degree analysis and multilevel analysis.

Out-degree analysis for analyzing personal networks

A personal network consists of respondent i in combination with his relations. The number of directed relations from this respondent is defined as his out-degree A_{i+}. Respondent i can have certain characteristics V_i such as age, gender, etc., and also his relations can have certain values W_{ij}, such as duration of the relation, frequency of contact, etc. Also the nominees in the personal network of respondent i can have certain values V_j, such as age, occupation etc. This way each respondent i together with his directed relations can be viewed as a star. The principle of out-degree analysis is to compute or estimate features of the total unknown network by using aggregated variables based on these personal networks (stars). One of the most basic aggregate variables of the personal network is the number of contacts single respondents have (out-degree A_{i+}). It is often relevant to classify the relations of vertex i according to categorical variables V_j referring to the vertices j nominated by i, or according to categorical variables W_{ij} referring to i's relations. The out-degree A_{i+} is then partitioned into a vector $(A_{i+1}, ..., A_{i+p})$ where A_{i+c} indicates the number of i's relations falling in class c. Subsequently, these aggregations can be analyzed with multivariate analysis techniques such as Hotelling's T^2-tests, principal component analysis. So, when using out-degree analysis, structure in a large, unknown network is described by means of independent respondents and their mentioned relations. This approach differs from the more standard methods of approaching network structure of complete networks in terms of graph-theoretic concepts such as density, cycles, and transitivity. Other features of the network can be indicated symbolically as aggregated sums or means:

$$O_i = \sum_{j=1}^{A_{i+}} f(V_i, V_j, W_{ij}) \quad or \quad O_i = \sum_{j=1}^{A_{i+}} \frac{f(V_i, V_j, W_{ij})}{A_{i+}}$$

where f is a (real or vector-valued) function of the attributes V_i of i, the attributes V_j of j, and the attributes W_{ij} of their directed relations

In the Rotterdam cocaine research project[5] respondents were asked to mention a maximum of 10 other cocaine users in the five pre-defined circuits; the world of entertainment (c1); at work (c2); at home or at private parties (intimate) (c3); at sport or hobbies clubs (c4); the hard drug scene (c5). These five circuits were defined with two purposes in mind: to 'force' the respondents also to nominate users in the periphery of his network (not only close friends) and to get insight into the contact patterns between and within the circuits. The known numbers of other users per circuit of each single respondent can be considered an indication of his opportunity structure concerning his possibilities to get in touch with cocaine. The five nominee circuits together with the respondent's circuit can be used to define meaningful variables (network features) aggregated from the personal network of each respondent. For each respondent a vector of five numbers is obtained; the number of known users per circuit. By averaging the number of nominees per circuit for each respondent's circuit, a 5*5 matrix is obtained. However, the field workers only found 4 and 1 respondent(s) respectively who were using cocaine mainly in the work circuit and in the hobby/sport circuit, so these respondent's circuits are left out of the analysis. What is left is a 3*5 matrix. The contact patterns over the five circuits can be analyzed by multivariate analysis of variance for determining if the three respondent's circuits differ concerning their distribution of the relations. The test statistic used was Wilk's ∧. By means of the Hotelling's T^2 two respondent's circuits can be analyzed to study if these two circuits differ significantly concerning their contact patterns over the five nominee circuits. Subsequently univariate t-statistics can be used to determine which of the nominee circuits contribute to the observed differences.

Data were collected by means of a combination of snowball sampling and targeted sampling. It is therefore worthwhile to compare the total snowball sample with the star sample. For meeting the assumptions of the star sampling design, respondents who are extensions of the initial respondents are left out of the sample population. For relaxing the assumption of the initial simple random sample, it has been checked if all respondents in the initial sample were indeed found independently of each other. This was not the case for all respondents, so it was decided to generate five subsamples in which, for every pair (or group) of respondents who were found dependently from each other by the fieldworkers, one was randomly selected to be in the star sample. These five samples are denoted star sample 1 to 5. In table 1 the generated 3*5 matrices are shown; PN-c1 are the personal networks of respondents in the entertainment circuit; PN-c3 are the personal networks of the respondents in the intimate circuit; PN-c5 are the personal networks of the respondents in the hard-drugs circuit; between brackets are the standard deviations.

Table 1   Average number of relations in the five circuits (in the columns) for respondents in the three circuits (rows)

Star sample 1

	c1	c2	c3	c4	c5
PN-c1 PN = 17	5.2 (4.3)	1.8 (3.2)	4.7 (2.9)	0.2 (1.0)	1.3 (3.3)
PN-c3 PN = 20	5.0 (4.0)	2.5 (3.4)	2.9 (3.7)	1.0 (2.1)	1.5 (2.2)
PN-c5 PN = 14	3.3 (3.2)	1.6 (2.0)	2.1 (2.5)	0.5 (1.1)	3.8 (4.2)

Star sample 2

	c1	c2	c3	c4	c5
PN-c1 PN = 17	5.7 (4.1)	2.0 (3.1)	4.4 (3.1)	0.4 (1.2)	1.3 (3.3)
PN-c3 PN = 19	4.7 (4.2)	2.6 (3.4)	3.2 (3.9)	1.1 (2.1)	1.5 (2.3)
PN-c5 PN = 15	3.9 (3.5)	2.5 (3.2)	1.6 (2.2)	0.7 (1.6)	4.9 (4.4)

Star sample 3

	c1	c2	c3	c4	c5
PN-c1 PN = 19	6.0 (4.0)	2.4 (3.5)	4.2 (3.5)	1.0 (2.0)	1.2 (3.1)
PN-c3 PN = 20	4.2 (4.1)	2.3 (3.4)	2.7 (3.8)	1.0 (2.1)	1.6 (2.2)
PN-c5 PN = 12	3.0 (2.6)	1.3 (1.6)	2.2 (2.7)	0.6 (1.2)	4.4 (4.2)

Star sample 4

	c1	c2	c3	c4	c5
PN-c1 PN = 17	5.4 (4.2)	1.9 (3.1)	4.5 (3.1)	0.5 (1.3)	1.4 (3.3)
PN-c3 PN = 20	4.9 (4.1)	2.5 (3.7)	2.9 (3.7)	1.0 (2.1)	1.6 (2.2)
PN-c5 PN = 14	3.8 (3.6)	1.9 (2.9)	1.6 (2.3)	0.7 (1.5)	4.9 (4.1)

Star sample 5

	c1	c2	c3	c4	c5
PN-c1 PN = 19	5.3 (4.1)	2.0 (3.0)	3.9 (3.2)	0.8 (1.9)	1.2 (3.1)
PN-c3 PN = 20	4.6 (4.1)	2.5 (3.4)	3.0 (3.8)	1.0 (2.1)	1.5 (2.2)
PN-c5 PN = 12	3.6 (3.3)	2.2 (3.0)	1.8 (2.4)	0.7 (1.6)	5.4 (4.2)

Average of star samples 1-5

	c1	c2	c3	c4	c5
PN-c1 PN = 19	5.5 (4.1)	2.0 (3.2)	4.3 (3.2)	0.6 (1.5)	1.3 (3.2)
PN-c3 PN = 20	4.7 (4.1)	2.5 (3.5)	2.9 (3.8)	1.0 (2.1)	1.5 (2.2)
PN-c5 PN = 12	3.5 (3.3)	1.9 (2.6)	1.9 (2.4)	0.6 (1.4)	4.7 (4.2)

Snowball sample

	c1	c2	c3	c4	c5
PN-c1 PN = 34	6.0 (3.9)	2.0 (3.0)	4.1 (3.3)	0.8 (1.7)	0.9 (2.3)
PN-c3 PN = 34	4.0 (3.7)	2.2 (3.0)	4.1 (4.0)	0.9 (1.9)	0.9 (1.8)
PN-c5 PN = 22	3.0 (3.3)	2.2 (2.9)	1.8 (2.2)	0.6 (1.3)	4.4 (4.1)

To be noted in table 1 are the high standard deviations indicating the large variation between the respondents in knowing other users per circuit. Another important aspect to be noted in table 1 are the differences in the nomination patterns between the star samples and the snowball sample concerning the respondents from the intimate circuit. The respondents in the star samples nominate systematically lower other users in the intimate circuit (c3) and systematically higher other users in the entertainment circuit (c1) and hard drug circuit (c5) than the respondents of the snowball sample. In other words those respondents (extensions) who are found via the initial respondents and using cocaine mostly at home or at private parties mention more users in their own circuit and less in the entertainment and hard drug circuit than the initial respondents. This phenomena can considered to be a 'snowball effect' due to the fact that most initial 'intimate circuit' respondents are not found in this circuit, but later assigned to it. These initial respondents are probably also using frequently in other circuits, so they have a higher probability of being found by fieldworkers in the zero stage of the snowball sample and subsequently those users who use mainly in the intimate circuit are under-represented. All matrices have significant Wilk's $\wedge$'s (α = .05), indicating that in each sample the three groups of personal networks differ concerning the distribution of their contact patterns over the five circuits (Table 2).

Table 2   Multivariate F-statistics for (PN-c1, PN-c3, PN-c5)

Sample	F-value	D.F.	Sig. of F
Star sample 1	2.08	(10, 88)	$p < .04$
Star sample 2	2.88	(10, 88)	$p < .01$
Star sample 3	2.53	(10, 88)	$p < .01$
Star sample 4	2.72	(10, 88)	$p < .01$
Star sample 5	3.18	(10, 88)	$p < .01$
Snowball sample	5.46	(10, 166)	$p < .01$

The univariate F-statistics, significant at α=.05, show which nominees circuits contribute to the multivariate significance of the differences between the contact patterns of three groups of personal networks (table 3).

Table 3   Significant univariate F-statistics for (PN-c1, PN-c3, PN-c5)

Sample	Circuit of nominees	F-value	D.F.	Sig. of F
Star sample 2	hard drugs	5.64	(2, 48)	$p < .01$
Star sample 3	hard drugs	4.45	(2, 48)	$p < .02$
Star sample 4	intimate	3.23	(2, 48)	$p < .05$
	hard drugs	6.08	(2, 48)	$p < .01$
Star sample 5	hard drugs	8.08	(2, 48)	$p < .01$
Snowball sample	entertainment	5.03	(2, 87)	$p < .01$
	intimate	3.79	(2, 87)	$p < .03$
	hard drugs	13.67	(2, 87)	$p < .01$

In all personal networks the numbers of known users in the hard drug circuit seem to be different in the three respondent circuits (except in star sample 1). In the snowball sample three nominee circuits contribute to the multivariate significance of differences between the contact patterns. This is not the case in the star populations. Note that the snowball sample is larger, leading to a higher power for the F tests, and that also there can be a systematic difference between the snowball sample and the star samples. However, all samples give a strong indication that the numbers of known users in the hard drug circuit, in particular, are different in the three respondent circuits.

In all six samples the contact patterns in the respondent circuits entertainment and intimate have more or less the same pattern. The Hotelling's T^2-tests for comparing the contact patterns between the personal networks of the entertainment and intimate circuit in all six samples show to be nonsignificant. In the five star samples no significant univariate differences are observed concerning the average known number of users per nominee circuit between respondents from the entertainment and intimate circuit. However, in the snowball sample a significant univariate difference between the respondents from the entertainment and intimate circuit is observed concerning the known number of users in the entertainment circuit; $F = 4.76$, D.F. $= (1, 62)$, $p < .04$ (due to the mentioned 'snowball effect'). All samples give strong indications that the contact patterns between respondents from the entertainment and intimate circuit are more or less the same, so the composition of the personal networks in the entertainment and intimate circuit can be regarded as identical and considered to be one circuit, in this paper denoted as C-circuit (combined circuit).

When comparing the relation patterns between respondents from the C-circuit and the hard drug circuit in all six samples a multivariate significant difference is observed (all $p < .01$). The univariate effects which contribute to the overall multivariate significance between respondents from the C-circuit and the hard drug circuit differ between the star samples and the snowball sample (table 4).

Table 4   Significant univariate F-statistics between C-circuit and PN-c5

Population	Circuit	F-value	D.F.	Sig. of F
Star sample 1	hard drugs	5.83	(1, 49)	p < .02
Star sample 2	intimate	4.68	(1, 49)	p < .04
	hard drugs	11.47	(1, 49)	p < .01
Star sample 3	hard drugs	8.94	(1, 49)	p < .01
Star sample 4	intimate	4.12	(1, 49)	p < .05
	hard drugs	12.37	(1, 49)	p < .01
Star sample 5	hard drugs	16.39	(1, 49)	p < .01
Snowball sample	entertainment	4.90	(1, 88)	p < .03
	intimate	7.05	(1, 88)	p < .01
	hard drugs	28.09	(1, 88)	p < .01

In the snowball sample the average known users in three circuits (entertainment, intimate and hard drug) differ between the personal networks of respondents from the C-circuit and the hard drug circuit, while in the star samples the average known users in the hard drug circuit contribute mainly to the difference in contact patterns between these two groups of respondents.

Concluding, the results of the out-degree analysis, based on the five star samples and the snowball sample show the existence of two different circuits in the network of cocaine users in Rotterdam; personal networks from the C-set and the hard drug circuit. In particular, the number of known users in the hard drug circuit is different in the composition of the personal networks between these two respondent circuits.

Multilevel analysis for analyzing personal networks

Personal network data have a hierarchical organisation in which the relations (level 1) are grouped within respondents (level 2). The rationale of multilevel analysis is the proposition that the hierarchical organisation of data should be taken into account in the analysis. In this appendix the two-level regression model or random coefficient model is briefly discussed, a more complete discussion of multilevel or random effects models can be found in Goldstein (1987) or Bryk & Raudenbush (1992).

The two-level regression model accounts for the fact that the respondent is a constant factor in his personal network and his relations are likely to have features in common. The model assumes that the intercepts and the regression coefficients per respondent may vary and be correlated because the respondents are considered to be a sample from a population of respondents.

Suppose all relationships obtained by star sampling have a score Y_{ij} where i is the relation and j the respondent and this score is predicted by two explanatory variables; X_{ij} at the relation-level with regression coefficient β_{10} and Z_j at the respondent-level with regression coefficient γ_{01}. The complete expression for the two-level regression model is (more explanatory variables at both levels can be added):

$$Y_{ij} = \beta_{00} + \gamma_{01}Z_j + \beta_{10}X_{ij} + \gamma_{11}Z_jX_{ij} + u_{0j} + u_{1j}X_{ij} + e_{ij}$$

where
β_{00} is the average score for the respondents;
γ_{11} is the effect Z_j has on X_{ij} (interaction effect);
u_{0j} is the unique effect of respondent j on the average score;
u_{1j} is the unique effect of respondent j on the effect of X;
e_{ij} is the residual at the relation-level.

This model takes into account the hierarchical structure of the data obtained by star sampling. In contrast to ordinary least squares regression this model has a more complex error term, i. e. $u_{0j} + u_{1j}X_{ij} + e_{ij}$. In the multilevel model the random errors do not need to be independent and have a constant variance for efficient estimation and accurate hypothesis testing. The model has random effects at two levels; a random effect u_{0j} at the respondent level, a random interaction effect $u_{1j}X_{ij}$ of the respondent with the variable at the relation-level, and a random effect e_{ij} at the relation level. In other words u_{0j} and u_{1j} are dependent 'within' each respondent because these components are common to every relation within respondent j. Another aspect of the complex error term $u_{0j} + u_{1j}X_{ij} + e_{ij}$ in the model is that this term depends upon u_{0j} and u_{1j}, which vary across the respondents, and upon the value of X_{ij} which varies across the relations, so the errors can have unequal variances. Several methods of estimation such as iterative generalized least squares method (IGLS) or restricted maximum such as lihood estimation (REML) can be used for estimating the parameters, but they are not discussed in this appendix (see Bryk and Raudenbush 1992).

To test whether the estimated values of the parameters are different from 0, indicating an effect is observable in the data, two methods must be distinguished. For testing single regression coefficients a simple t-test is sufficient

$$T(\theta) = \hat{\theta}/SE(\hat{\theta})$$

in which Θ is the regression coefficient.

For testing the components of variance or sets of regression coefficients, the principle of the likelihood ratio test can be used in multilevel analysis. By computing the deviance (minus twice the log likelihood), which can be considered a measure of distance between the model and the data, two models are tested. In the model eight free parameters can be observed $(r_1=8)^{(6)}$. The deviance of this model, denoted by $D(M_1)$, is computed. Depending on how many variance components are tested, 1 or more parameters are fixed, and for this model M_0 with r_0 free parameters the deviance $D(M_0)$ is computed. For testing if the given random effect exists in the model, the following hypotheses are tested:

H_0: M_0 is appropriate
H_1: M_1 is appropriate, but M_0 is not.
The test statistic is $D(M_0)-D(M_1)$ which, if H_0 holds, is χ^2 distributed with $(r_1 - r_0)$ degrees of freedom.

In multilevel models the percentage explained variance R^2 can be computed as the relative decrease in the mean squared errors with respect to the empty model, that is, the model which has, as random terms, only the random intercepts e_{ij} and u_{0j}. For a more detailed discussion see Snijders & Bosker (1992). The percentage explained variance R^2 is specified for two levels:

- for level 1, the relations,

$$R^2_{Y_{ij}} = 1 - \frac{\sigma^2 + \tau_{00}(A)}{\sigma^2 + \tau_{00}(0)}$$

in which (A) indexes the model with covariates and (0) the empty model

- for level 2, the respondents

$$R^2_{Y_j} = 1 - \frac{\dfrac{\sigma^2}{n} + \tau_{00}(A)}{\dfrac{\sigma^2}{n} + \tau_{00}(0)}$$

in which (A) the model with covariates, (0) the empty model and n the size of the personal network.

The out-degree analysis resulted in two main circuits of contact patterns concerning the spread of relations over the five circuits; personal networks in the C-circuit and in the hard drug circuit. Using a two-level regression analysis we want to study another aspect of the opportunity structure, namely the significance of cocaine in the relations compared between the personal networks from the C-circuit to the personal networks from the hard drug circuit. In this way we can explore how the significance of cocaine in the relationships in the C-circuit is to the relationships in the hard drug circuit. For the analyses the whole snowball sample is used. To detect if there any snowball effects observable, respondents of star sample 1 form the initial sample[7], while the other respondents are considered as extensions. The resulting data set used in the analysis contains 1124 relations within 88 respondents (51 initial respondents, 37 extensions).

The dependent variable Y_{ij} roughly measures the significance of cocaine of respondent j in relation i (see diagram 7.1). The following variables are used to describe the composition of the personal networks concerning the significance of cocaine:

At level 1 (relations)
Circuit of main use of the nominee (C-nom)   1 hard drug circuit
    2 sport/hobby circuit
    3 intimate circuit
    4 work circuit
    5 entertainment circuit

At level 2 (respondents)
Circuit of main use of the respondent (C-res) 1 hard drug circuit
                                                      2 sport/hobby circuit
                                                      3 intimate circuit
                                                      4 work circuit
                                                      5 entertainment circuit

The extension variable   0 no extension
                                                         1 respondent is extension

The amount of variance at the two levels is (between brackets standard errors)

Model I (no explanatory variables)

random residual at level 1	$\sigma^2 = 0.65$ (0.03)
random residual at level 2	$\tau_{00} = 0.41$ (0.07)
deviance	Dev = 2888.9

The intra-respondent correlation is 0.39 indicating the proportion of the total variance between respondents due to variation between the respondents (if this correlation is 0, a single level model can be used), so there is a fair amount of homogeneity of relations within respondents. Both levels have little variance because almost 50% of the Y_{ij}'s have value 2, so the assumption of normality is problematic. However, it is worthwhile to explore if the circuits of main use for respondents and nominees can reduce a part of the variance in model I.

In this appendix we briefly describe the followed procedure (for a detailed description of all steps and resulting models see Spreen 1992b). The first variable added to the model is the extension variable; to examine if this variable has a main effect concerning the significance of cocaine. It is assumed that the importance of cocaine in the extensions within the same circuit might be similar because they are likely to be from the same social environment. No significant main effect is observed (regression coefficient -0.12, standard error 0.16). Subsequently the categorical variables C-res and C-nom are added as sets of dummy variables to model I. Because we are interested in differences between the circuits within the C-circuit with respect to the hard drug circuit, the reference group of the respondent circuits (C-res) are the respondents of the hard drug circuit. The reference group of the nominee circuits (C-nom) is the hard drug circuit. Subsequently the eight possible interaction effects of the extension variable with the circuits of the C-set of both levels are added. The next step for describing the composition of the personal networks is to examine whether there are interactions between the respondent's and nominees' circuits (cross-level interactions). A last step is to examine if there are random coefficients. The final model is model II (the significant main effects are printed boldface; the sums of the dummy variables are given because they indicate the number of relationships in the indicated category, so it can be checked if an effect has a substantive meaning).

The decrease in deviance ($\chi^2_d = 186.5$; df = 11; p < .001) is clearly significant. The constant in model II is the average score of Y_{ij} from the respondents in the hard drug circuit with respect to nominees in the hard drug circuit. All dummy variables have a negative sign indicating that for the respondents from the C-circuit the importance of cocaine in their relations is lower compared to respondents from the hard drug circuit. Respondents from the hobby/sport and intimate circuits have a significant difference from respondents of the hard drug circuit indicating that the significance of cocaine in these personal networks is lower compared to the personal networks in the hard drug circuit. However, in the hobby/sport circuit there is only 1 respondent, who mentions 9 relations all having score $Y_{ij} = 1$, so this main effect has no substantive meaning, but for illustrative purposes this respondent (circuit) is given.

Model II

constant	2.85 (0.18)	
Level 2 C-res		Sum of dummies
respondents sport/hobby circuit	**-0.94 (0.15)**	9
respondents intimate circuit	**-0.32 (0.14)**	386
respondents work circuit	-0.21 (0.16)	56
respondents entertainment	-0.06 (0.12)	439
extensions	-0.33 (0.19)	368
Level 1 C-nom		
nominees sport/hobby circuit	**-1.72 (0.89)**	69
nominees intimate circuit	**-0.71 (0.22)**	290
nominees work circuit	-0.34 (0.45)	178
nominees entertainment circuit	**-0.60 (0.22)**	431
Interaction extensions by respondents intimate circuit	**0.47 (0.21)**	139
Interaction respondent sport/hobby circuit and nominees entertainment circuit	**0.67 (0.26)**	4
random residual at level 1	σ^2 = 0.46 (0.02)	
random residual at level 2	τ_{00} = 0.45 (0.09)	
random coefficient respondent work circuit	**0.60 (0.19)**	
random coefficient respondent intimate circuit	**0.24 (0.10)**	
random coefficient respondent entertainment circuit	**0.41 (0.11)**	
deviance	Dev = 2702.4	

Relations in which nominees mainly use cocaine in the sport/hobby, intimate or entertainment circuits have a significant lower importance of cocaine compared to relations in which nominees mainly use cocaine in the hard-drugs circuit.

The interaction of the extension variable with respondents in the intimate circuit shows to be significant. For all respondents of the intimate circuit the importance of cocaine in their relations is lower than for respondents of the hard drug circuit. However, the importance of cocaine is higher for respondents in the first snowball wave than for those in the initial sample. This may be due to a combination of the earlier discussed 'snowball effect' and chaining processes by assuming that the co-operative 'extension respondents' are cocaine users who are more open users and less reluctant to disclose their use. These extensions form a homogeneous group by having a higher importance of cocaine in their relations. This effect may be exaggerated due to capitalization on chance because it is the only significant one among eight tested interaction effects, but the significant effect is rather strong (regression coefficient 0.47, standard error 0.21), so the effect is added to model II.

Only one cross-level interaction is significant which is probably due to capitalization on chance (16 cross-level interactions were added to the model). The observed cross-level effect is an interaction between the single respondent from the sport/hobby circuit and four nominees from the entertainment circuit, so the meaning of this effect is negligible. However, for illustrative purposes this interaction effect is added

The significance of the random coefficients means that the respondents in the work, intimate and entertainment circuits are more different from each other concerning the significance of cocaine than those in the hard drug circuit, so adding these regression coefficients as fixed effects would give inaccurate estimations. Therefore the significance of cocaine in the relations of these circuits can considered to be very variable.

Concluding remarks

The results of the introduced link-tracing methodology show that this combination potentially provides a practical way for describing aspects of network structure in hidden populations. Out-degree analysis is used to examine the spread of relations across the pre-defined users circuits, while multilevel analysis is used to examine other aspects of this spread of relations. In particular, the comparison between the star samples and the snowball samples can be used to make well-founded inferences. Based on the out-degree analysis we may conclude that there are two circuits in Rotterdam. The multilevel analysis shows that the significance of cocaine in the relationships of respondents in the C-circuit is less than for respondents in the hard drug circuit. The multilevel analysis also shows that the C-circuit is not a homogeneous group. Respondents from the work, intimate, and entertainment circuits show a larger variability among their relations concerning the significance of cocaine than those from the hard-drugs circuit.

The observed 'snowball' effect in the intimate circuit shows that the 'intimate circuit' respondents can be better reached in the first wave of the sample, because these respondents are actually found in that circuit (more in the core of the intimate circuit). However, the chaining processes observed in the multilevel analysis make it harder to obtain well-founded inferences about these respondents, because the respondents in the first wave are more likely to be found due to their open use and co-operative behaviour.

Acknowledgements

The author is indebted to Tom Snijders, Holly Raider, Bert Bieleman, Cilia ten Den, Edgar de Bie, and Charles Kaplan for their helpful comments.

Notes

1. The author is presently working as a research fellow at the ICS, Department of Sociology, University of Groningen on the project 'Designs for the estimation of network parameters'. During the cocaine project he was working as assistent at Intraval, Groningen-Rotterdam and the Department of Statistics and Measurement, University of Groningen
2. For a discussion on the nature and the definition of hidden populations see Spreen (1992b) in which the distinction between egocentric and sociocentric hidden populations is described. Egocentric hidden populations are those populations in which the variable that defines its members is socially sensitive or subject to taboo but will not lead to connections between the members (for instance persons who commit child abuse). Sociocentric hidden populations also are characterized by a 'sensitive' variable, but the meaning of this variable has a social nature in the sense that contact patterns between members are presupposed. This appendix deals with sociocentric hidden populations.
3. Kruskal and Mosteller define chunk sampling as collecting whatever data is readily available.
4. In this appendix the in-degree of an individual in a network is defined as the total number of other individuals who would state that they have a relation with that individual; the out-degree is defined as the total number of other individuals an individual mentions as his relations.
5. The analysis in this appendix concerns only Rotterdam.
6. These eight free parameters are: intercept β_{00}, regression coefficients β_{10}, γ_{01}, γ_{11}, $\tau_{00} = \text{var } u_{0j}$, $\tau_{11} = \text{var } u_{1j}$, and $\tau_{01} = \text{cov } (u_{0j}, u_{1j})$.
7. Star sample 1 is arbitrarily chosen.

Lines across Europe

References

Biernacki, P., D. Waldorf (1981):
Snowball sampling: Problems and techniques of chain referral sampling. Sociological Methods and Research, 10, 2, 141-163.

Bryk, A.S., S. Raudenbush (1992):
Hierarchical linear models for social and behavioral research: applications and data analysis method. Sage Publications, Newbury Park.

Capobianco, M., O. Frank (1982):
Comparison of statistical graph-size estimators. Journal of Statistical Planning and Inference, 6, 87-97.

Coleman, J.S. (1958):
Relational Analysis: The study of social organizations with survey methods. Human Organization, 17, 4, 28-36.

Erickson, B.H. (1978):
Some problems of inference from chain data. In: K.F. Schuessler (ed.): Sociological Methodology. Josey-Bass, London, 276-302.

Frank, O. (1977):
Statistical inference in graphs. Journal of Statistical Planning and Inference, 1, 235-264.

Goldstein, H. (1987):
Multilevel models in educational and social research. Oxford University Press, New York.

Goodman, L.A. (1961):
Snowball sampling. Annals of Mathematical Statistics, 32, 148-170.

Horvitz, D.G., D.J. Thompson (1952):
A generalization of sampling without replacement from a finite universe. Journal of the American Statistical Association, 47, 663-685.

Kish, L. (1987):
Statistical design for research. Wiley & Sons, New York.

Klovdahl, A.S. (1989):
Urban social networks: Some methodological problems and possibilities. In: M. Kochen (ed.), The Small World. Norwood, New Jersey.

Kruskal, W., F. Mosteller (1979-80):
Representative sampling, I, II, III, and IV. International Statistical Review, 47, 48.

Marsden, P.V. (1990):
Network data and measurement. Annual Review of Sociology, 16, 435-463.

Snijders, T.A.B. (1992):
Estimation on the basis of snowball samples: how to weight? Bulletin de Methodologie Sociologique, 36, 59-70.

Snijders, T.A.B., R.J. Bosker (1992):
Modeled variance in two-level models. To be published.

Spreen, M. (1992a):

Rare populations, hidden populations, and link-tracing designs: what and why? Bulletin de Methologie Sociologique, 36, 34-58.

Spreen, M. (1992b):

Star sampling, out-degree analysis, and multilevel analysis: A practical link-tracing methodology for sampling and analyzing sociocentric hidden populations. Department of Sociology, University of Groningen, Groningen.

APPENDIX D

TABLES CHARACTERISTICS

1-GENDER	BARCELONA		ROTTERDAM		TURIN	
	% Respondents	% Nominees	% Respondents	% Nominees	% Respondents	% Nominees
Male	60	69	81	77	67	67
Female	40	31	19	23	33	33
Total (%)	100%	100%	100%	100%	100%	100%
Total (N)	153	434	110	1047	100	147

2-AGE	BARCELONA		ROTTERDAM		TURIN	
	% Respondents	% Nominees	% Respondents	% Nominees	% Respondents	% Nominees
Up to 15	0	1	0	0	0	0
from 16 to 20	4	4	10	9	8	3
from 21 to 25	27	31	21	24	18	19
from 26 to 30	34	34	26	28	46	37
from 31 to 35	22	20	21	18	11	28
from 36 to 40	8	7	18	14	10	9
from 41 to 50	5	3	4	6	7	2
over 50	0	0	0	1	0	2
Total (%)	100%	100%	100%	100%	100%	100%
Total (N)	153	434	110	1042	100	147

3-DAILY WORK	BARCELONA		ROTTERDAM		TURIN	
	% Respondents	% Nominees	% Respondents	% Nominees	% Respondents	% Nominees
Employed	75	86	39	not available	70	79
Unemployed	5	3	40		17	8
Student	11	11	7		10	9
Other*	9	0	14		3	3
Total (%)	100%	100%	100%		100%	100%
Total (N)	153	434	110		100	150

* Other: Housewives, persons in prison, persons living in closed institutions, etcetera.

4-OCCUPATION	BARCELONA		ROTTERDAM		TURIN	
	% Respondents	% Nominees	% Respondents	% Nominees	% Respondents	% Nominees
Professional	13	10	13	not available	3	18
Intermediate	32	25	20		20	20
Skilled	49	58	59		61	43
Unskilled	6	7	8		16	19
Total (%)	100%	100%	100%		100%	100%
Total (N)	114	340	39		70	102

5-PROFESSION	BARCELONA		ROTTERDAM		TURIN	
	% Respondents	% Nominees	% Respondents	% Nominees	% Respondents	% Nominees
Professional	8	8	8	5	6	not available
Arts/Culture	16	14	17	19	9	
Care	5	3	8	13	3	
Education	1	2	5	2	5	
Sales	8	19	4	5	22	
Media	17	9	1	2	3	
Pubs etcetera	10	12	19	19	8	
Technical	18	18	27	23	38	
Administrative	19	14	5	7	8	
Sex-industrie	0	1	6	5	0	
Total (%)	100%	100%	100%	100%	100%	
Total (N)	102	327	78	606	64	

6-PRESENT USE	BARCELONA		ROTTERDAM		TURIN	
	% Respondents	% Nominees	% Respondents	% Nominees	% Respondents	% Nominees
Still using	83	92	65	91	74	92
Not using	17	8	35	9	26	8
Total (%)	100%	100%	100%	100%	100%	100%
Total (N)	153	404	110	400	100	140

7-EDUCATIONAL LEVEL	BARCELONA	ROTTERDAM	TURIN
	% Respondents	% Respondents	% Respondents
Primary school	22	0	33
Secondary school	52	68	51
University	26	32	16
Total (%)	100%	100%	100%
Total (N)	145	109	99

Primary school: 4 - 5 years to 12 - 14 years
Secondary school: 12 - 14 years to 17 - 20 years
University: professional or university education (± 17 - 20 years to 21 - 25 years)

There are some differences between the cities in the way the educational level of the respondents has been classified. Whereas Barcelona and Turin looked upon the degree of the respondents, Rotterdam looked upon the highest level of education the respondent attended. The respondent may or may not have obtained a degree at that level.

8-AGE AT INITIATION	BARCELONA	ROTTERDAM	TURIN
	% Respondents	% Respondents	% Respondents
Up to 15	5	18	0
from 16 to 20	56	44	57
from 21 to 25	26	25	30
from 26 to 30	9	9	11
from 31 to 35	3	3	1
from 36 to 40	2	1	1
Total (%)	100%	100%	100%
Total (N)	149	110	100

9-LENGHT OF USE	BARCELONA	ROTTERDAM	TURIN
	% Respondents	% Respondents	% Respondents
< 1 year	0	7	6
1 - 2	13	8	12
3 - 4	9	16	10
5 - 7	29	24	21
8 - 10	21	22	29
> 10	27	22	22
Total (%)	100%	100%	100%
Total (N)	149	106	100

10-KIND OF DRUG	BARCELONA	ROTTERDAM	TURIN
	% Respondents	% Respondents	% Respondents
Only cocaine	0	2	4
Also others drugs	81	65	55
Also heroin	19	33	41
Total (%)	100%	100%	100%
Total (N)	151	110	98

11-COCAINE USE	BARCELONA	ROTTERDAM	TURIN
	% Respondents	% Respondents	% Respondents
Continuous	95	79	77
Discontinuous	5	21	23
Total (%)	100%	100%	100%
Total (N)	153	101	100

12-PATTERN OF USE	BARCELONA	ROTTERDAM	TURIN
	% Respondents	% Respondents	% Respondents
Decreasing	5	3	1
Increasing	12	34	12
Equal	24	12	16
Peak	47	34	17
Twin peaks	7	12	26
Discontinuous	5	5	28
Total (%)	100%	100%	100%
Total (N)	153	110	92

13a--WAY OF USE DURING FIRST PERIOD			
HEROIN USERS	BARCELONA	ROTTERDAM	TURIN
	% Respondents	% Respondents	% Respondents
Sniffing	82	50	80
Injecting	18	27	18
Basing	0	6	0
Other*	0	17	2
Total (%)	100%	100%	100%
Total (N)	28	35	39

* Other: Smoking in a cigarette and chasing the dragon

13b--WAY OF USE DURING FIRST PERIOD			
NON HEROIN USERS	**BARCELONA**	**ROTTERDAM**	**TURIN**
	% Respondents	% Respondents	% Respondents
Sniffing	96	98	100
Injecting	3	0	0
Basing	0	1	0
Other*		1	0
Total (%)	100%	100%	100%
Total (N)	125	74	59

* Other: Smoking in a cigarette and chasing the dragon

14a--WAY OF USE DURING LAST PERIOD			
HEROIN USERS	**BARCELONA**	**ROTTERDAM**	**TURIN**
	% Respondents	% Respondents	% Respondents
Sniffing	50	22	85
Injecting	46	33	15
Basing	4	17	0
Other*	0	28	0
Total (%)	100%	100%	100%
Total (N)	28	35	40

* Other: Smoking in a cigarette and chasing the dragon

14b--WAY OF USE DURING LAST PERIOD			
NON HEROIN USERS	**BARCELONA**	**ROTTERDAM**	**TURIN**
	% Respondents	% Respondents	% Respondents
Sniffing	98	89	100
Injecting	0	0	0
Basing	2	10	0
Other*	0	0	0
Total (%)	100%	100%	100%
Total (N)	125	74	59

* Other: Smoking in a cigarette and chasing the dragon

15a--WAY OF USE DURING HEAVIEST PERIOD			
HEROIN USERS	**BARCELONA**	**ROTTERDAM**	**TURIN**
	% Respondents	% Respondents	% Respondents
Sniffing	32	17	74
Injecting	64	49	26
Basing	4	17	0
Other*	0	17	0
Total (%)	100%	100%	100%
Total (N)	28	36	39

* Other: Smoking in a cigarette and chasing the dragon

15b--WAY OF USE DURING HEAVIEST PERIOD			
NON HEROIN USERS	**BARCELONA**	**ROTTERDAM**	**TURIN**
	% Respondents	% Respondents	% Respondents
Sniffing	97	86	98
Injecting	0	0	0
Basing	3	10	0
Other*	0	4	2
Total (%)	100%	100%	100%
Total (N)	125	74	60

* Other: Smoking in a cigarette and chasing the dragon

16--CIRCUIT OF USE DURING FIRST PERIOD			
	BARCELONA	**ROTTERDAM**	**TURIN**
	% Respondents	% Respondents	% Respondents
Entertainment	46	27	11
Work	10	5	3
Intimate	37	47	83
Hobby/sports	1	2	0
Hard drugs	6	18	3
3 equally important circuits	0	0	0
Total (%)	100%	100%	100%
Total (N)	144	110	100

17--CIRCUIT OF USE DURING LAST PERIOD			
	BARCELONA	**ROTTERDAM**	**TURIN**
	% Respondents	% Respondents	% Respondents
Entertainment	69	35	32
Work	5	4	4
Intimate	6	35	59
Hobby/sports	0	1	0
Hard drugs	7	25	5
3 equally important circuits	13	0	0
Total (%)	100%	100%	100%
Total (N)	144	107	98

18--CIRCUIT OF USE DURING HEAVIEST PERIOD			
	BARCELONA	**ROTTERDAM**	**TURIN**
	% Respondents	% Respondents	% Respondents
Entertainment	62	29	36
Work	8	6	6
Intimate	2	33	52
Hobby/sports	0	1	0
Hard drugs	8	30	6
3 equally important circuits	20	0	0
Total (%)	100%	100%	100%
Total (N)	141	106	99

19a--FREQUENCY OF USE DURING FIRST PERIOD			
HEROIN USERS	**BARCELONA**	**ROTTERDAM**	**TURIN**
	% Respondents	% Respondents	% Respondents
Daily	25	56	15
Weekly	7	29	40
Monthly	18	12	43
Sporadic	50	3	2
Total (%)	100%	100%	100%
Total (N)	28	34	40

19b--FREQUENCY OF USE DURING FIRST PERIOD			
NON HEROIN USERS	**BARCELONA**	**ROTTERDAM**	**TURIN**
	% Respondents	% Respondents	% Respondents
Daily	8	8	10
Weekly	14	30	27
Monthly	10	35	48
Sporadic	68	27	15
Total (%)	100%	100%	100%
Total (N)	124	74	60

20a--FREQUENCY OF USE DURING LAST PERIOD			
HEROIN USERS	**BARCELONA**	**ROTTERDAM**	**TURIN**
	% Respondents	% Respondents	% Respondents
Daily	36	75	17
Weekly	21	8	33
Monthly	21	11	38
Sporadic	21	6	12
Total (%)	100%	100%	100%
Total (N)	28	36	40

20b--FREQUENCY OF USE DURING LAST PERIOD			
NON HEROIN USERS	**BARCELONA**	**ROTTERDAM**	**TURIN**
	% Respondents	% Respondents	% Respondents
Daily	14	19	10
Weekly	24	28	28
Monthly	25	35	50
Sporadic	37	18	12
Total (%)	100%	100%	100%
Total (N)	124	74	59

21a--FREQUENCY OF USE DURING HEAVIEST PERIOD			
HEROIN USERS	**BARCELONA**	**ROTTERDAM**	**TURIN**
	% Respondents	% Respondents	% Respondents
Daily	82	86	49
Weekly	7	8	36
Monthly	7	3	15
Sporadic	4	3	0
Total (%)	100%	100%	100%
Total (N)	28	36	39

21b--FREQ. OF USE DURING HEAVIEST PERIOD			
NON HEROIN USERS	BARCELONA	ROTTERDAM	TURIN
	% Respondents	% Respondents	% Respondents
Daily	43	38	27
Weekly	32	38	34
Monthly	10	17	34
Sporadic	15	7	6
Total (%)	100%	100%	100%
Total (N)	124	74	59

22a--LEVEL OF USE DURING FIRST PERIOD			
HEROIN USERS	BARCELONA	ROTTERDAM	TURIN
	% Respondents	% Respondents	% Respondents
< 0.5	18	17	0
0.5 - 1.00	20	5	3
1 -- 2.5	36		31
2.5 -- 5	18	11	16
5 -- 10	4	22	29
10 -- 15	0	29	0
> 15	4	17	21
Total (%)	100%	100%	100%
Total (N)	28	23	38

22b--LEVEL OF USE DURING FIRST PERIOD			
NON HEROIN USERS	BARCELONA	ROTTERDAM	TURIN
	% Respondents	% Respondents	% Respondents
< 0.5	52	41	15
0.5 - 1.00	27	19	24
1 -- 2.5	10	4	18
2.5 -- 5	4	10	7
5 -- 10	4	14	17
10 -- 15	2	4	0
> 15	1	8	18
Total (%)	100%	100%	100%
Total (N)	124	51	46

Tables characteristics

23a--LEVEL OF USE DURING LAST PERIOD			
HEROIN USERS	**BARCELONA**	**ROTTERDAM**	**TURIN**
	% Respondents	% Respondents	% Respondents
< 0.5	18	8	10
0.5 - 1.00	11	0	3
1 – 2.5	25	8	26
2.5 – 5	14	4	10
5 – 10	11	12	24
10 – 15	0	40	0
> 15	21	28	26
Total (%)	100%	100%	100%
Total (N)	28	25	27

23b--LEVEL OF USE DURING LAST PERIOD			
NON HEROIN USERS	**BARCELONA**	**ROTTERDAM**	**TURIN**
	% Respondents	% Respondents	% Respondents
< 0.5	42	26	14
0.5 - 1.00	17	17	21
1 – 2.5	21	12	23
2.5 – 5	10	9	10
5 – 10	6	14	16
10 – 15	2	15	0
> 15	2	7	16
Total (%)	100%	100%	100%
Total (N)	124	74	40

24a--LEVEL OF USE DURING HEAVIEST PERIOD			
HEROIN USERS	**BARCELONA**	**ROTTERDAM**	**TURIN**
	% Respondents	% Respondents	% Respondents
< 0.5	0	4	0
0.5 - 1.00	0	4	0
1 – 2.5	4	0	16
2.5 – 5	11	0	5
5 – 10	14	7	19
10 – 15	32	63	0
> 15	39	22	59
Total (%)	100%	100%	100%
Total (N)	28	36	37

24b--LEVEL OF USE DURING HEAVIEST PERIOD			
NON HEROIN USERS	BARCELONA	ROTTERDAM	TURIN
	% Respondents	% Respondents	% Respondents
< 0.5	12	15	6
0.5 - 1.00	12	7	9
1 -- 2.5	10	5	23
2.5 -- 5	16	7	11
5 -- 10	18	20	13
10 -- 15	19	24	2
> 15	13	22	36
Total (%)	100%	100%	100%
Total (N)	124	74	53

25--INCOME DURING FIRST PERIOD			
	BARCELONA	ROTTERDAM	TURIN
	% Respondents	% Respondents	% Respondents
Legal	not available	53	67
Semi - legal		19	13
Criminal		28	17
Total (%)		100%	100%
Total (N)		110	98

26--INCOME DURING LAST PERIOD			
	BARCELONA	ROTTERDAM	TURIN
	% Respondents	% Respondents	% Respondents
Legal	73	47	68
Semi - legal	not available	27	14
Criminal	27	26	17
Total (%)	100%	100%	100%
Total (N)	145	110	98

27--INCOME DURING HEAVIEST PERIOD			
	BARCELONA	ROTTERDAM	TURIN
	% Respondents	% Respondents	% Respondents
Legal	not available	39	61
Semi - legal		18	14
Criminal		43	24
Total (%)		100%	100%
Total (N)		110	98

Tables characteristics

28a--CRIMINAL ACTIVITIES DURING FIRST PERIOD			
HEROIN USERS	**BARCELONA**	**ROTTERDAM**	**TURIN**
	% Respondents	% Respondents	% Respondents
Yes	39	69	47
No	61	31	53
Total (%)	100%	100%	100%
Total (N)	28	35	40

28b--CRIMINAL ACTIVITIES DURING FIRST PERIOD			
NON HEROIN USERS	**BARCELONA**	**ROTTERDAM**	**TURIN**
	% Respondents	% Respondents	% Respondents
Yes	2	19	8
No	98	81	92
Total (%)	100%	100%	100%
Total (N)	125	74	60

29a--CRIMINAL ACTIVITIES DURING LAST PERIOD			
HEROIN USERS	**BARCELONA**	**ROTTERDAM**	**TURIN**
	% Respondents	% Respondents	% Respondents
Yes	46	57	50
No	54	43	50
Total (%)	100%	100%	100%
Total (N)	28	35	40

29b--CRIMINAL ACTIVITIES DURING LAST PERIOD			
NON HEROIN USERS	**BARCELONA**	**ROTTERDAM**	**TURIN**
	% Respondents	% Respondents	% Respondents
Yes	8	15	7
No	92	85	93
Total (%)	100%	100%	100%
Total (N)	124	74	60

30a--CRIMINAL ACTIVITIES DURING HEAVIEST PERIOD			
HEROIN USERS	BARCELONA	ROTTERDAM	TURIN
	% Respondents	% Respondents	% Respondents
Yes	61	83	72
No	39	17	28
Total (%)	100%	100%	100%
Total (N)	28	36	40

30b--CRIMINAL ACTIVITIES DURING HEAVIEST PERIOD			
NON HEROIN USERS	BARCELONA	ROTTERDAM	TURIN
	% Respondents	% Respondents	% Respondents
Yes	10	31	10
No	90	69	90
Total (%)	100%	100%	100%
Total (N)	125	74	60

31--INVOLVEMENT IN THE MARKET			
	BARCELONA	ROTTERDAM	TURIN
	% Respondents	% Respondents	% Respondents
Yes	28	32	36
No	72	68	64
Total (%)	100%	100%	100%
Total (N)	153	108	98

32--MOST IMPORTANT CIRCUIT OF OBTAINMENT			
	BARCELONA	ROTTERDAM	TURIN
	% Respondents	% Respondents	% Respondents
Entertainment	3	8	1
Intimate	41	27	43
Home dealer, no heroin	35	30	19
Home dealer, + heroin	3	24	15
Market	15	11	22
>3 equally important circuits	4	0	0
Total (%)	100%	100%	100%
Total (N)	144	99	100

Tables characteristics

33a--COCAINE RELATED PROBLEMS			
HEROIN USERS	BARCELONA	ROTTERDAM	TURIN
	% Respondents	% Respondents	% Respondents
Yes	93	83	38
No	7	17	62
Total (%)	100%	100%	100%
Total (N)	28	36	40

33b--COCAINE RELATED PROBLEMS			
NON HEROIN USERS	BARCELONA	ROTTERDAM	TURIN
	% Respondents	% Respondents	% Respondents
Yes	71	40	63
No	29	60	37
Total (%)	100%	100%	100%
Total (N)	125	74	60

34a--KIND OF COCAINE RELATED PROBLEMS			
HEROIN USERS	BARCELONA	ROTTERDAM	TURIN
	% Respondents	% Respondents	% Respondents
Physical	38	0	36
Psychological	12	23	20
Social	0	6	12
Economic	0	10	8
>3 different problems	50	61	24
Total (%)	100%	100%	100%
Total (N)	26	31	15

34b--KIND OF COCAINE RELATED PROBLEMS			
NON HEROIN USERS	BARCELONA	ROTTERDAM	TURIN
	% Respondents	% Respondents	% Respondents
Physical	58	7	36
Psychological	23	40	45
Social	0	17	0
Economic	8	10	14
>3 different problems	11	26	5
Total (%)	100%	100%	100%
Total (N)	89	30	38

APPENDIX E

PATTERNS OF USE

1 - Decreasing

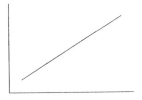

2 - Increasing

3 - Same level

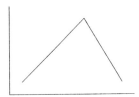

4 - Peak

5 - Twin peaks

6 - Discontinuous